1 MONTH OF
FREE
READING

at

www.ForgottenBooks.com

By purchasing this book you are eligible for one month membership to ForgottenBooks.com, giving you unlimited access to our entire collection of over 1,000,000 titles via our web site and mobile apps.

To claim your free month visit: www.forgottenbooks.com/free443431

ISBN 978-0-484-24597-5
PIBN 10443431

Production Note

Cornell University Library produced this volume to replace the irreparably deteriorated original. It was scanned using Xerox software and equipment at 600 dots per inch resolution and compressed prior to storage using CCITT Group 4 compression. The digital data were used to create Cornell's replacement volume on paper that meets the ANSI Standard Z39.48-1984. The production of this volume was supported in part by the Commission on Preservation and Access and the Xerox Corporation. 1990.

A

PRESENTATION

OF THE

THEORY OF HERMITE'S FORM

OF

LAMÉ'S EQUATION

WITH A DETERMINATION OF THE EXPLICIT FORMS IN TERMS OF
THE p FUNCTION FOR THE CASE n EQUAL TO THREE.

— ·· ` —

CANDIDATES THESIS

FOR THE

DEGREE OF DOCTOR OF PHILOSOPHY

PRESENTED BY

J. BRACE CHITTENDEN, A. M.,

PARKER FELLOW OF HARVARD UNIV., INSTRUCTOR IN PRINCETON COLLEGE.

TO THE

PHILOSOPHICAL FACULTY OF THE ALBERTUS-UNIVERSITÄT

OF

KÖNIGSBERG IN PR.

PRINTED BY B. G. TEUBNER, LEIPZIG.

1893.

DEDICATED

TO THE FIRST OF MY MANY TEACHERS,

MY

MOTHER

WHO MORE THAN ALL OTHERS HAS RENDERED THE REALI-
ZATIONS OF MY STUDENT LIFE POSSIBLE, FOR WHOM NO
SACRIFICE HAS BEEN TO GREAT IN FURTHERING THE
INTERESTS OF HER SONS.

Introduction.

The following thesis is practically a presentation of the general analytical theory of Lamé's differential equation of the form known as Hermite's. The underlying principles and also the general solutions are therefore necessarily based upon the original work of M. Hermite, published for the first time in Paris in 1877 in the *Comptes rendus* under the title *"Sur quelques applications des fonctions elliptiques"* and on a later treatment of the subject by Halphen in his work entitled *"Traité des Fonctions Elliptiques et leur applications"*, Vol. II, Paris 1888. M. Hermite has employed the older Jacobian functions while Halphen has used in every case the Weierstrass p function, and not only the notation but the ultimate forms as well as the complex functions in which they are expressed are in the two works intirely different.

As far as I know, no attempt has before been made to establish the absolute relations of these different functions. In attempting to do this, I have developed the intire theory in a new presentation, working out the results of M. Hermite in terms of the p function, having principly in view a determination of the explicit values of all the forms for the special case n equal to three.

I may add that owing to the exceptional privilege granted by the Minister of Education and the Philosophical Faculty of the Albertus-Universität allowing the publishing of this thesis in English,

I am not without hope that this general presentation of the theory of Lamé's Functions may prove a welcome addition to the literature of the subject where in English Todhunter's "Lamé's and Bessel's Functions" is the only representative. Finally I must acknowledge my indebtedness to Prof. Lindemann not only for the direction of a most valuable course of reading but for a general although, owing to a lack of time, a by no means detailed review of the work.

Contents.

Thesis.

Part I.

Historical Development and Definition of the Equation of Lamé.

The Problem of Lamé.

In order to arrive at an understanding of the highly generalized forms that have taken the name of Lamé it is adivisable to return for the moment to the original problem of the potential in which they claim a common origin.

Lagrange and Laplace (1782) in their researches with respect to the earth regarded as a solid sphere developed the potential function*) which led to the development of the theory of the Kugelfunction. From this date until 1839 the only name that need be mentioned is that of Fourier (1822) who, in developing his theory of heat solved the problem with reference to a right angled cylinder discovering the series named after him.

In the following decade**) however Lamé***) generalized the work of his predicessors by solving the problem for an ellipsoid with three unequal axes thus laying the foundation for the development of functions of which the former are but special cases. He used to this end the inductive method arriving at special solutions through a study of the problem already solved with reference to the sphere.

The problem of Lamé may be stated thus:

Let the surface of an ellipsoid be given by the equation $u = u_0$; it is required to find a function T which will satisfy the equation of the potential and which for the value $u = u_0$ will reduce to a given

---

*) See note Heine, Handbuch der Kugelfunctionen, p. 2, Berlin 1878, and Heine, 2ᵈ vol. Zusätze zum ersten Bande.

**) See also reference to Green Heine p. 1.

***) Memoire sur les axes des surfaces isothermes du second degree considérés comme des fonctions de la température. Journal des Mathématiques pures et appliqués. 1ʳᵉ série. t. IV, p. 103. 1839.

function of v and w, where T is the temperature at a point whose elliptic coordinates are u, v and w.

The working eliments are then, the potential function, generally written

[1] $\quad \cdot \quad \cdot \quad \cdot \quad \cdot \quad \cdot \quad \cdot \quad \cdot \quad \cdot \quad \cdot \quad \sum_{\alpha} \dfrac{d^2 u}{dx_\alpha^2} = 0$

or transformed in terms of the p function

[2] $\quad \cdot \quad \cdot \quad (pv - pu)\dfrac{d^2 T}{du^2} + (pu - pv)\dfrac{d^2 T}{dv^2} + (pu - pv)\dfrac{d^2 T}{dw^2} = 0$

the relation,

[3] $\quad \cdot \quad \cdot \quad \cdot \quad \cdot \quad \cdot \quad \cdot \quad \cdot \quad T = f(u)\, f(v)\, f(w)$

and the equation

[4] $\quad \cdot \quad \cdot \quad \cdot \quad \cdot \quad \cdot \quad \cdot \quad \cdot \quad \dfrac{d^2 y}{du^2} = [A p u + B] y$

where $y = f(u)$ and A and B are constants.

If T is developed by Maclaurin's theorem with respect to the rectangular coordinates, we may write: *)

[5] $\quad \cdot \quad \cdot \quad \cdot \quad \cdot \quad T = T_0 + T_1 + T_2 + \cdots + T_n + \cdots$

where T_n in general is an intire homogenious polynomial of the nth degree, it is observed that each of the functions T_n will also satisfy [1], the equation of the potential, in which case [1] would be an intire homogeneous polynomial of the $(n - 2)^d$ degree. This polynomial must be identically zero which will impose $\frac{1}{2}(n - 1)n$ linear conditions. The quantities T_n will have in all $\frac{1}{2}(n + 1)(n + 2)$ constants, which leaves the difference $2n + 1$ equal to the number of constants that may be considered arbitrary.

Now the general expression for x^2 in terms of p is known to be

[6] $\quad \cdot \quad \cdot \quad \cdot \quad \tau^2 x_\alpha^2 = \dfrac{(pu - e_\alpha)(pv - e_\alpha)(pw - e_\alpha)}{(e_\alpha - e_\beta)(e_\alpha - e_\gamma)}$

τ being a constant, from which we see that by a change of variable T_n may become an intire homogeneous function of the nth degree with respect to the variables

[7] $\quad \cdot \quad \cdot \quad \cdot \quad \cdot \quad \sqrt{pu - e_1}, \quad \sqrt{pu - e_2}, \quad \sqrt{pu - e_3}$

quantities proportional to the axes of the ellipsoid, and of the 1st degree, pu being of the second and $p'u$ of the third.

We have then that T, the function sought, is composed of similar functions T_n, where T_n is of the nth degree, is symmetrical

---

*) see Halphen. Vol. II p. 466.

with respect to u, v and w and having $2n + 1$ arbitrary constants, is capable of satisfying the equation [2] of the potential.

From the above relations we derive

[8] $\cdots\quad \dfrac{d^2 T}{du^2} = f''u f v f w = T\dfrac{f''u}{fu} = [Apu + B]T$

with corresponding equations for v and w.

If then one can find $2n + 1$ systems of constants A and B of such sort that for each of these systems there exists a solution $y = fu$ of equation [4] where y is an intire function of the nth degree each of the corresponding products $fu\,fv\,fw$ will furnish a term T_n of T and the problem of Lamé will be solved. The value of A for all of these systems is $n(n + 1)$ where n may be considered as always positive, since the substitution $n \sim -(n + 1)$ does not alter the value of A.

The Problem of Hermite.

Continuing our review we find that one of the original forms of Lamé's equation expressed in terms of the Jacobian function is

[9] $\cdots\quad \dfrac{d^2 y}{dx^2} - [n(n + 1)k^2 sn^2 x + h]y = 0$

corresponding to the form [4]*)

[10] $\cdots\cdots\quad \dfrac{d^2 y}{du^2} - [n(n + 1)pu + B]y = 0$

where h is an arbitrary constant and n a positive whole number.

Lamé succeeded in finding the requisite number of values of h to complete his solution for the ellipsoid and the solutions of [4] corresponding to these values are known as the original special functions of Lamé.

The problem then arose:

Required to determine a solution of Lamé's original equation which shall hold for any values of h and n.

Except for the special values $n = 1$ and $n = 2$ no advance was made towards a solution until M. Hermite**), making use of the progress in the theory of functions inaugurated by Cauchy, arrived at the solution and by so doing opened a new field for

---

*) See transformation p. 20.

**) Sur quelques applications des fonctions elliptiques. Comptes rendus de l'académie des sciences de Paris. 1877.

‚the application of the elliptic functions and leading later to the integration of a large class of differential equations.*)

In this connection M. Hermite introduces the functions called by him doubly periodic of the second species, which have the special property, that save for a constant factor they remain unaltered upon the addition to the argument of the fundimental periods.

The solution of M. Hermite developed in terms of snu and for n odd may be ‚written in the form

$$[11] \cdot \cdot \ y = F(u) = \frac{D_u^{2\nu-1}f(u)}{\Gamma^{(2\nu)}} + \frac{h_1\,D_u^{2\nu-3}f(u)}{\Gamma^{(2\nu-2}} + \cdots + h_{\nu-1}f(u)$$

where $n = 2v - 1$, with a corresponding form for n even, where $f(u)$ is a doubly periodic function of the second species, namely,

$$f(u) = e^{\lambda(\alpha-ik')}\chi(u)$$

where

$$\chi(u) = \frac{H'(0)\,H(u+\omega)}{\Theta(u)\,\Theta(\omega)}\; e^{-\frac{\Theta'(\omega)}{\Theta(\omega)}(u-iK')+\frac{i\pi\omega}{2K}}$$

That this shall be a solution the quantities ω and λ must be determined to correspond with definite conditions and herein lies the chief difficulty when explicit values of the functions are sought. Moreover the above development fails as we shall find when seeking to deduce the special functions of M. Mittag-Leffler from the general form.

M. Hermite was thus led to a new presentation of the general solution in the form of a product, namely

$$y = \prod_{a=a\cdot b\cdots}\frac{\sigma(u+a)}{\sigma a\,\sigma u}\;e^{-u\zeta a}$$

a form of solution suited to every case.

The general theory based upon the latter solution has been lately perfected by Halphen**), who, confining himself in the main to the use of the p function, presents the subject in an excellent but highly condensed form.

---

*) Equations of M. Éimile Picard. Comptes rendus, t. XC, p. 128 and 293. — Prof. Fuchs, Ueber eine Classe von Differenzialgleichungen, welche durch Abelsche oder elliptische Functionen integrirbar sind. Nachrichten von Göttingen 1878, and Hermite: Annali di Matematica, serie II, Bd. IX, 1878.
**) Traité des Fonctions Elliptiques et leur applications. B. II. Paris 1888.

Definitions.

Returning to form [9] of Lamé's equation we observe that it has the following properties:

It has a coefficient

$$n(n + 1)k^2 sn^2 x + h$$

that is doubly periodic and has only one infinite $x = iK'$ and its congruents, and it is known to have an integral which is a rational function of the variable. Conforming with these peculiarities M. Mittag-Leffler*) defines *the general Hermite's form of Lamé's equation of the nth order as a linear homogenious differential equation of the order n having coefficients that are doubly periodic functions, having the fundimental periods $2K$ and $2iK'$ and everywhere finite save in the point $x = iK'$ and its congruents which alone are infinite and whose general integral is a rational function of the variable.*

The general theory of Herrn Fuchs**) then gives the form, namely

[12] · · · · · · $Y^n + \Phi_2(x)y^{(n-2)} + \cdot\cdot + \Phi_n(x)y = 0$

where

$$\Phi_2(x) = \alpha_0 + \alpha_1 sn^2 x$$
$$\Phi_3(x) = \beta_0 + \beta_1 sn^2 x + \beta_2 D_x^2 sn^2 x$$
$$\Phi_4(x) = \gamma_0 + \gamma_1 sn^2 x + \gamma_2 D_x sn^2 x + \gamma_3 D_x^2 sn^2 x$$

— — — — — — — — — —
— — — — — — — — — —

But a better generalization based upon a full representation of the singular points is given by Prof. Klein***) and later stated as follows by Dr. Bôcher†). First the ordinary form of the equation of Lamé may through transformation become††)

[13] $\dfrac{d^2y}{dx^2} + \dfrac{1}{2}\left(\dfrac{1}{x - e_1} + \dfrac{1}{x - e_2} + \dfrac{1}{x - e_3}\right)\dfrac{dy}{dx} = \dfrac{Ax + B}{4(x - e_1)(x - e_2)(x - e_3)} y$

where the exponents of the zeros e_1, e_2, e_3 are 0 and $\frac{1}{2}$ and that of the infinites $\dfrac{1 \pm \sqrt{1 + 4A}}{4}$. From this generalizing by the introduction of n zeros we have the following definition:

— — — — — — — —

*) Annali di Matematica, tomo XI, 1882.
**) Comptes rendus etc. 1880. p. 64.
***) Math. Annal. Bd. 38.
†) Ueber die Reihentwickelungen der Potentialtheorie. Göttingen 1891.
††) See also transformation p. 20.

„*Mit dem Namen Lamésche Gleichung bezeichnen wir eine überall regulüre homogene Differentialgleichung zweiter Ordnung mit rationalen Coefficienten, deren im Endlichen gelegene singuläre Punkte e_1, $e_2 \cdot \cdot e^n$ sämmtlich die Exponenten 0, $\frac{1}{2}$ besetzen und in unendlichen nur einen uneigentlich singularen Punkt aufweist.*‘

Lamé's equation becomes in accordance with this definition and freed from the possibility of a logarithmic irrationality through a determination of the coefficient of x^{n-3}:

$$[14] \qquad \qquad \frac{d^2 y}{dx^2} + \frac{f'(x)}{2fx}\frac{dy}{dx}$$

$$- \frac{1}{4f(x)}\left[\frac{-n(n-4)}{4}x^{n-2} + \frac{(n-2)(n-4)}{4}\sum_1^n E_i\, x^{n-3} + A^{n-4} + \cdots + M\right] y = 0$$

where

$$f(x) = (x - e_1)(x - e_2)\cdots(x - e_n).$$

It ·is further evident that this form, like the Hermite form and as previously developed by Prof. Heine, is but a special case of a general equation of a higher order.

In speaking of Lamé's equation we will understand an equation conforming with the above definition whose general form is given by [14] and, if the order is higher than the second, distinguish by mentioning the order.

Forms [9] and [10] will then be called Hermite's forms of Lamé's equation or simply Hermite's equation, where again the order need be mentioned only if it be other than the second.

Any solution of any form of Lamé's equation will be a *function of Lamé* and if the doubly periodic functions first determined by Lamé are mentioned they will be designated as the *special functions of Lamé.*

Part II.

Hermite's Integral as a Sum.

The Function of the Second Species.

We have the problem *required the integral y of the equation*

[15] $\quad \cdots \quad \cdots \quad \cdots \quad \dfrac{d^2 y}{d u^2} = [n(n+1)k^2 sn^2 u + h]\, y$

where h is taken arbitrarily, n is any intire positive number and k is the modulus of the elliptic function.

M. Hermite introduces to this end a function which he names *doubly periodic* of the *second species*, which may be defined as *a product of a quotient composed of σ functions, the number of zeros being equal to the number of the infinites, and an exponential, having the property of reproducing itself multiplied by an exponential factor when the variable is increased by the periods 2K and 2iK.*

It is defined then in general by the relations:

$$F(u + 2K) = \mu\, F(u)$$

[16] $\qquad\qquad F(u + 2iK') = \mu'\, F(u)$

$$F(u) = \frac{\sigma(u - a_1)\,\sigma(u - a_2)\ldots\sigma(u - a_{n-1})}{\sigma(u - h_1)\,\sigma(u - h_2)\ldots\sigma(u - b_{n-1})}\, e^{\varrho u}.$$

The factors μ and μ' are called *Multiplicators.*

M. Hermite might have been led to the employment of this function by the following analysis which is essentially that given by Halphen.*)

Consider for the moment that y be such a function of the second species but having instead of the n different poles but one pole $u \equiv o$ of the nth order in which case the function will have n roots. Upon developing the properties of this function one finds that its second derivative has the same multiplicator as the function

---

*) Bd. II p. 495.

itself and that therefore the quotient $y'' : y$ will not only be doubly periodic but will have a single pole $u \equiv o$ of the second order.

This function then satisfies the necessary conditions and the corresponding quotient $\dfrac{y''}{y}$ may then be written equal to

$$n(n + 1)\, sn^2\, x + h$$

where h is a constant. But we have taken this function with the condition that it have but one pole of the order n subject to the above conditions which affords n arbitrary constants and employing also an arbitrary constant factor we obtain $(n + 1)$ arbitraries in all. That is sufficient to satisfy all the conditions and leave h to be chosen at will. Hence we must conclude that there is no reason why y should not be a doubly periodic function of the second species and our problem reduces to the determination of a function · whose general form and properties [16] are known.

From this standpoint we have:

Required a function such that

$$F(u + \Omega) = \mu F(u), \qquad \Omega = 2K$$
$$F(u + \Omega') = \mu' F(u), \qquad \Omega' = 2iK'.$$

Define:

[17] · · · · · · · $f(u) = A\, \dfrac{\sigma(u + v)}{\sigma(u)}\, e^{\lambda u}$ *)

which function we will speak of as the *Eliment* the general form [16] being a product of similar eliments.

We have the fundimental relations:

$$\sigma(u + \Omega) = - \sigma(u) e^{2\eta u + \eta \Omega}$$
$$\sigma(u + \Omega') = - \sigma(u) e^{2\eta' u + \eta' \Omega'}$$
$$\eta = \frac{\sigma'}{\sigma}\left(\frac{\Omega}{2}\right) = \zeta\left(\frac{\Omega}{2}\right).$$

Whence

$$f(u + \Omega) = A\, \frac{\sigma(u + v)}{\sigma(u)}\, e^{\lambda(u + \Omega) + 2\eta v}$$
$$= f(u)\varrho^{\lambda \Omega + 2\eta v}.$$

Choosing then v and λ correctly we may write

$$\mu = e^{\lambda \Omega + 2\eta v}$$

---

*) Hermite, in the following analysis, employs the function given on p. 11, namely the function χ expressed in terms of the Θ function.

with a corresponding value for μ' and we may then write

$$\frac{F(u)}{f(u)} = \Phi(u)$$

where Φ is a doubly periodic function, that is

$$\Phi(u + m\Omega + n\Omega') = \Phi(u).$$

Again

$$f(u - \Omega) = \frac{1}{\mu} f(u) \text{ and } f(u - \Omega') = \frac{1}{\mu} f(u).$$

Whence

$$f(u - z - \Omega) = \frac{1}{\mu} f(u - z) \text{ where } F(z + \Omega) = \mu F(z)$$

and we derive

[18] $\cdots \quad \cdots \quad \cdots \quad \Phi(z) = F(z) f(u - z)$

where Φ is doubly periodic.

From this point the development of $F(u)$ depends upon the theory of Cauchy, as it is obtained by calculating the residuals of Φ for the values of the argument that render it infinite and equating the sum to zero as follows.

First $f(u)$ becomes infinite for the value $u \equiv 0$ whence its residual

$$E_{u=0} f(u) = [uf u]_{u=0} = A \frac{[\sigma(u + v) e^{\lambda u}]_{u=0}}{\left[\frac{\sigma u}{u}\right]_{u=0}} = A \frac{\sigma(v)}{\sigma'(0)} = A \sigma(v)$$

and becomes equal to unity if we take

$$A = \frac{1}{\sigma(v)}.$$

Whence

[19] $\cdots \quad \cdots \quad \cdots \quad f(u) = \frac{\sigma(u + v)}{\sigma(u) \sigma(v)} e^{\lambda u}.$

Again

$$E_u \Phi(z) = \lim_{z = u} (z - u) \Phi(z) = \lim_{z = u} (z - u) F(z) f(u - z)$$

and developing $f(u - z)$ we have

$$E_u \Phi(z) = - F(u)$$

Again let a be any pole of $F(u)$ in which case, developing by the function theory, we may write

$$F(a + \varepsilon)_{\varepsilon = 0} = A \varepsilon^{-1} + A_1 D_\varepsilon \varepsilon^{-1} + A_2 D_\varepsilon^2 \varepsilon^{-1} + \cdots$$
$$+ A_\alpha D_\varepsilon^\alpha \varepsilon^{-1} + a_0 + a_1 \varepsilon + a_2 \varepsilon^2 + \cdots$$

2*

and

$$f(u - a - \varepsilon) = f(u - a) - \frac{\varepsilon}{1} D_u f(u - a) + \frac{\varepsilon^2}{1 \cdot 2} D_u^2 f(u - a) - \cdots$$

$$+ \frac{(-1)^\alpha \varepsilon^\alpha}{1 \cdot 2 \cdots \alpha} D_u^\alpha (fu - a) + \cdots$$

where

$$D_\varepsilon^n \varepsilon^{-1} = (-1)^n \frac{1 \cdot 2 \cdots n}{\varepsilon^{n+1}}$$

We have then

$$E_a \Phi = \lim_{\varepsilon = 0} \varepsilon F(a + \varepsilon) f(u - a - \varepsilon)$$

$$= A f(u - a) + A_1 D_u f(u - a) + A_2 D_u^2 f(u - a) + \cdots$$

$$+ A_\alpha D_u^\alpha f(u - a)$$

with similar expressions for E_b, E_c . . .

But Φ being a doubly periodic function we know that the sum of its residuals with respect to u, a, b . . equals zero whence

[20] $$F(u) = \sum_{a = a, b, c..} [A f(u - a) + A_1 D_u f(u - a) + A_2 D_u^2 f(u - a) + \cdots$$

$$+ A_\alpha D_u^\alpha f(u - a)]$$

where A_i is determined from the first development.

This important formula still further narrows our problem to a consideration of $f(u)$ in terms of which and its derivatives under conditions to be determined it is now evident that $y = F_1(u)$ may be expressed.

Transformation of Hermite's Equation.

We have written Hermite's equation in its original form

[21] · · · · · · $$\frac{d^2 y}{dx^2} = [n(n + 1) k^2 s n^2 x + h].$$

That this is however but a special case of a more general form is seen as follows.

Take the integral

$$x = \int_0^\lambda \frac{d\lambda}{\sqrt{(1 - \lambda^2)(1 - k^2 \lambda^2)}} = \int_0^\lambda \frac{d\lambda}{\sqrt{A}}.$$

We have

$$\frac{dy}{d\lambda} = \frac{dy}{dx} \cdot \frac{dx}{d\lambda} = \frac{dy}{dx} \cdot \frac{1}{\sqrt{A}}$$

or

$$\frac{dy}{dx} = \sqrt{A} \frac{dy}{d\lambda}$$

whence

$$\frac{d^2y}{d\lambda^2} = \frac{d^2y}{dx^2} \cdot \frac{1}{\varDelta} - \frac{1}{2}\frac{\varDelta'}{\varDelta^{\frac{3}{2}}}\frac{dy}{dx}$$

or

$$\frac{d^2y}{dx^2} = \varDelta\frac{d^2y}{d\lambda^2} + \frac{1}{2}\varDelta'\frac{dy}{d\lambda}.$$

Substituting we derive the ordinary form of Lamé's equation

[22] . . . . $\varDelta\frac{d^2y}{d\lambda^2} + \frac{1}{2}\varDelta'\frac{dy}{d\lambda} - [n(n+1)k^2sn^2x + h] = 0.$*)

The value of $\varDelta$ gives as singular points ± 1; $\pm\frac{1}{k}$ and ∞.

For our present purpose however we need the equation expressed in terms of u and pu which is derived from (21) by means of the relations

$$p(u) = e_3 + \frac{e_1 - e_3}{sn^2u\sqrt{e_1 - e_3}}, \quad k^2sn^2(u+ik') = \frac{1}{sn^2u}$$

and making the substitutions:

$$x = u\sqrt{e_1 - e_3} \quad u \sim u + ik'$$
$$dx^2 = du^2(e_1 - e_3)$$

we obtain:

$$\frac{d^2y}{du^2} \cdot \frac{1}{(e_1 - e_3)} = [n(n+1)\frac{pu - e_3}{e_1 - e_3} + h].$$

Define

[23] . . . . $B = h(e_1 - e_3) - n(n+1)e_3.$

Whence our equation may be written:

[24] . . . . $y'' = [n(n+1)pu + B]y.$

Development of the Integral.

We observe, since snx reduces to zero only for the value $x = 0$, that we have but one pole of the second order in Hermite's equation and that we may therefore develop y within a cercle whose radius is less than Ω', the form being

$$y = u^\nu[y_0 + y_1u + y_2u^2 + \cdots]$$

whence

$$y' = \nu u^{\nu-1}y_0 + (\nu+1)y_1u^\nu + (\nu+2)u^{\nu+1}y_2 + (\nu+3)u^{\nu+2}y_3 + \cdots$$
$$y'' = \nu(\nu-1)u^{\nu-2}y_0 + \nu(\nu+1)y_1u^{\nu-1} + (\nu+1)(\nu+2)u^\nu y_2 + \cdots$$

---

*) Compair general form [14] p. 16.

We have also

$$p(u) = \frac{1}{u^2} + c_1 u^2 + c_2 u^4 + \cdots$$

These values in [24] give:

$$\nu(\nu - 1) y_0 u^{\nu-2} + \cdots = n(n+1) y_0 u^{\nu-2} + \cdots$$

whence:

$$\nu(\nu - 1) = n(n+1) \quad \text{or} \quad \nu = -n.$$

This value gives since the uneven powers fall out

[25] $\cdots \quad y = \dfrac{1}{u^n} + \dfrac{h_1}{u^{n-2}} + \dfrac{h_2}{u^{n-4}} + \cdots + \dfrac{h_i}{u^{n-2i}} + \cdots$

from which we again derive

$$y'' = n(n+1)\frac{1}{u^{n+2}} + (n-2)(n-1)\frac{h_1}{u^n} + (n-4)(n-3)\frac{h_2}{u^{n-2}}$$

$$+ (n-6)(n-5)\frac{h_3}{u^{(n-4)}} + \cdots + (n-2i)(n-2i+1)\frac{h_i}{u^{n-2i+2}}$$

$$= \left[\frac{1}{u^n} + \frac{h_1}{u^{n-2}} + \frac{h_2}{u^{n-4}} + \cdots + \frac{h_i}{u^{n-2i}} + \cdot \right]$$

$$\left[n(n+1)\left(\frac{1}{u^2} + c_1 u^2 + c_2 u^4 + \cdots\right) + B \right]$$

$$= n(n+1)\frac{1}{u^{n+2}} + \frac{h_1}{u^n}(n+1)n + \cdots + \frac{h_i n(n+1)}{u^{n-2i+2}} + \cdots$$

$$+ B\frac{1}{u^n} + \frac{B h_1}{u^{n-2}} + \frac{B h_2}{u^{n-4}} + \cdots + \frac{B h_i}{u^{n-2i}} + \cdots$$

$\bullet \quad + n(n+1) c_1 \dfrac{1}{u^{n-2}} + n(n+1) c_1 h_1 \dfrac{1}{u^{n-4}} + \cdots$

$$+ n(n+1) c_1 h_i \frac{1}{u^{n-2i+2}} + \cdots + n(n+1) c_2 \frac{1}{u^{n-4}} + \cdots$$

$$+ \,-\,-\,-\,-\,-\,-\,-$$

Equating the coefficients of like powers of u in this identity one finds

$$\frac{1}{u^n} \mid (n-2)(n-1) h_1 = n(n+1) h_1 + B$$

$$\frac{1}{u^{n-2}} \mid (n-4)(n-3) h_2 = n(n+1)[h_2 + c_1] + h_1 B$$

$$\frac{1}{u^{n-4}} \mid (n-6)(n-5) h_3 = n(n+1)(h_3 + c_1 h_1 + c_2) + h_2 B$$

$$\frac{1}{u^{n-6}} \mid (n-8)(n-7)h_4 = n(n+1)(h_4 + c_1 h_2 + c_2 h_1 + c_3) + h_3 B$$

$$- \quad - \quad - \quad - \quad - \quad - \quad - \quad - \quad -$$

$$\frac{1}{u^{n-2i+2}} \mid (n-2i)(n-2i+1)h_i = n(n+1)[h_i + c_1 h_{i-2} + c_2 h_{i-3} + \cdots$$
$$+ c_{i-1}] + h_{i-1} B.$$

Whence we obtain all the coefficients in the development for y by means of the recurring formula.

[26] $$2i(2i - 2n - 1)h_i = n(n+1)[c_1 h_{i-2} + c_2 h_{i-3} + \cdots$$
$$+ c_{i-2} h_i + c_{i-1}] + h_{i-1} B.$$

Since then h_i is determined we have when n is even and equal to 2ν

$$y = \frac{1}{u^{2\nu}} + \frac{h_1}{u^{2\nu-2}} + \cdots + \frac{h_{\nu-1}}{u^2} + h_\nu$$

and if n be odd and equal to $2\nu - 1$

$$y = \frac{1}{u^{2\nu-1}} + \frac{h_1}{u^{2\nu-2}} + \cdots + \frac{h_{\nu-1}}{u} + h_\nu u$$

where h_i is given by (26).

Development of the Eliment $f(u)$.

Having now a development of y we can, if we develop $f(u)$ and substitute in the development of $F(u)$, find by comparison the conditions necessary that $y = F_1(u)$ be a solution.

We have then to determine the development of

$$f(u) = \frac{\sigma(u+v)}{\sigma(u)\,\sigma(v)} e^{\lambda u}$$

$$= \frac{1}{u} + \cdots$$

since $\qquad [uf(u)]_{u=0} = 1$

To this end we develop first the function

[27] $\quad \cdot \quad \cdot \quad \cdot \quad \cdot \quad$ $$\varphi(u) = f(u)e^{(\lambda+\zeta v)u} = \frac{\sigma(u+v)}{\sigma(u)\,\sigma(v)} e^{-u\zeta v}$$

We have:

$$pu = -\frac{d\zeta u}{du} = -\frac{d}{du} \cdot \frac{\varrho' u}{\varrho u} = \frac{1}{u^2} + c_1 u^2 + c_2 u^4 + \cdots$$

whence

$$\frac{\sigma' u}{\sigma u} = \frac{1}{u} - \frac{1}{3} c_1 u^3 - \frac{1}{5} c_2 u^5 - \frac{1}{7} c_3 u^7 \cdots$$

By Taylor's theorem:

$$\frac{\sigma'}{\sigma}(u+v) = \frac{\sigma'}{\sigma}(v) + u\,\frac{d\,\frac{\sigma'}{\sigma}(v)}{du} + \frac{u^2}{1.2}\,\frac{d^2\,\frac{\sigma'}{\sigma}(v)}{du^2} + \cdots$$

$$= \zeta(v) - u\,p(v) - \frac{u^2}{1.2}\,p'(v) - \frac{u^3}{1.2.3}\,p''(v)\cdots$$

Passing now to logarithms we derive:

$$\frac{\varphi'}{\varphi}(u) = \frac{\sigma'}{\sigma}(u+v) - \frac{\sigma'}{\sigma}(u) - \frac{\sigma'}{\sigma}(v)$$

$$= -\frac{1}{u} - u\,p(v) - \frac{u^2}{2}\,p'(v) - \frac{u^3}{3}\left(\frac{p''(v)}{3!} - \frac{c_1}{3}\right)$$

$$-\frac{u^4}{4!}\,p'''(v) - \frac{u^5}{5!}\left[p''''(v) - \frac{c_2}{5}\right] - \cdots$$

$$= -\frac{1}{u} + A_2 u + \frac{A_3}{2!}\,u^2 + \frac{A_4}{3!}\,u^3 + \cdots$$

Integrating we have:

$$\log\varphi = -\log u + A_2\frac{u^2}{2!} + A_3\frac{u^3}{3!} + A_4\frac{u^4}{4!} + \cdots$$

whence

[28] $$\varphi = \frac{1}{u}\,e^{\left[A_2\frac{u^2}{2!} + A_3\frac{u^3}{3!} + \cdots\right]}$$

$$= \frac{1}{u}\left[1 + A_2\frac{u^2}{2!} + A_3\frac{u^3}{3!} + \cdots\right] + \frac{1}{u}\left[A_2\frac{u^2}{2!} + A_3\frac{u^3}{3!} + \cdots\right]^2 + \cdots$$

$$= \frac{1}{u}\left[1 + P_2\frac{u^2}{2!} + P_3\frac{u^3}{3!} + P_4\frac{u^4}{4!} + \cdots\right]$$

where

$$P_2 = A_2 = -p(v); \quad P_3 = A_3 = -p'(v);$$

$$P_4 = -3p^2(v) + \frac{3}{5}g_2 = A_4 + 3A_2^2$$

$$P_5 = -3pvp'v = A_5 + 10A_2A_3 \quad \text{etc.}$$

showing that the coefficients P_i are intire functions of pv and $p'v$.*)

---

*) The functions P_i correspond to the functions Ω in Hermite's treatis, for example

$$P_2 = -p(v) = \Omega = \varkappa^2 sn^2 u - \frac{1+\varkappa^2}{3}$$

$$P_3 = -p'u = \Omega_1 = \varkappa^2 snu\,cnu\,dnu$$

see p. 126 development of χ.

From these forms we pass immediately to

[29] $f(u) = \varphi(u) e^{(\lambda + \zeta v)u}$

$$= \varphi(u)\left[1 + (\lambda + \zeta v) + (\lambda + \zeta v)^2 \frac{u^2}{1 \cdot 2} + \cdots\right]$$

$$= \frac{1}{u}\left\{\left[1 + (\lambda + \zeta u) u + (P_2 + (\lambda + \zeta u)^2) \frac{u^2}{1 \cdot 2}\right]\right.$$

$$\left. + \left[P_3 + 3 P_2 (\lambda + \zeta u) + (\lambda + \zeta u)^3\right] \frac{u^3}{1 \cdot 2 \cdot 3} + \cdots\right\}$$

$$= \frac{1}{u} + H_0 + H_1 u + H_2 u^2 + H_3 u^3 + \cdots$$

Take

$$\lambda = x - \zeta v$$

whence

[30] · · $f(u) = \frac{\sigma(u+v)}{\sigma(u)\,\sigma(v)} e^{(x - \zeta v) u}$

$$= \frac{1}{u} + x + (x^2 + P_2) \frac{u}{2} + (x^3 + 3 P_2 x + P_3) \frac{u^2}{2 \cdot 3}$$

$$+ (x^4 + 6 P_2 x^2 + 4 P_3 x + P_4) \frac{u^3}{2 \cdot 3 \cdot 4} + \cdots$$

$$= \frac{1}{u} + H_0 + H_1 (u) + H_2 (u)^2 + H_3 u^3.$$

Where in Hermite's Notation

$$H_0 = x.$$

$$H_1 = \tfrac{1}{2} (x^2 + P_2)$$

[31] $\qquad H_2 = \tfrac{1}{6} (x^3 + 3 P_2 x + P_3)$

$$H_3 = \tfrac{1}{24} (x^4 + 6 P_2 x^2 + 4 P_3 x + P_4)$$

— — — — — — — — —

Determination of the Integral.

We are now enabled to determine the exact expression for $F(u)$ and the conditions necessary that it become equal to y by a process of comparison of the several developments obtained.

First we have:

$$f(u) \;=\; \frac{1}{u} + H_0 + H_1 u + H_2 u^2 + \cdots + H_i u^i + \cdots$$

$$f'(u) \;=\; -\frac{1}{u^2} + H_1 + 2 H_2 u + 3 H_2 u^2 + \cdots + i H_i u^{i-1} + \cdots$$

$$f''(u) \;=\; +\frac{2}{u^2} + 2 H_2 + 2 \cdot 3 H_3 u + \cdots + i(i-1) H_i u^{i-2}$$

$$f'''(u) \;=\; -\frac{2 \cdot 3}{u^4} + 2 \cdot 3 H_3 + \cdots$$
$$+ i(i-1)(i-2) H_i u^{i-3} + \cdots$$

— — — — — — — — — — — — — — — — —

$$f^{(n-1)}_{(n \text{ odd})} \;=\; + \frac{2 \cdot 3 \cdots (n-1)}{u^n} + 2 \cdot 3 \cdots (n-1) H_{n-1} + \cdots$$
$$+ i(i-1) \cdots (i-n+1) H_i u^{i-n+1} + \cdots$$

Again

$$y_{n=2\nu-1} = \frac{1}{u^{2\nu-1}} + \frac{h_1}{u^{2\nu-3}} + \cdots + \frac{h_{\nu-1}}{u} + h_\nu u$$

$$y_{n=2\nu} = \frac{1}{u^{2\nu}} + \frac{h_1}{u^{2\nu-2}} + \cdots + \frac{h_{\nu-1}}{u^2} + h_\nu.$$

And in general

$$y = F_1 u = A_\alpha f^{(\alpha)} + A_{\alpha-1} f^{(\alpha-1)} + \cdots + f$$
$$= A_\alpha f^{(n-1)} + A_{\alpha-2} f^{(n-3)} + \cdots + f \quad (n \text{ odd}).$$

Now substituting the values $f^{(x)}$ found above and ordering the coefficients so that the residual with respect to u will be unity we find by comparison that we may write

[32] $\quad \cdot \; y = F_1(u) = \dfrac{1}{(n-1)!} f^{(n-1)} + \dfrac{1}{(n-3)!} h_1 f^{(n-3)} + \cdots h_{\nu-1} f$

$\hfill$ (n odd and $=2\nu-1$)

provided x and ν be so taken that the constant term equal zero and the coefficient of the next term equal h_ν and

[33] $y = F_2(u) = - \dfrac{1}{(n-1)!} f^{(n-1)} - \dfrac{h_1}{(n-3)!} f^{(n-3)} - \dfrac{h_2}{(n-5)!} f^{(n-5)} - \cdots$

$\hfill - h_{\nu-1} f' \quad$ (n even and $= 2\nu$)

provided x and ν be so taken that the constant term equal h_ν and the coefficient of the next term equal zero
or in general

[34] $(-1)^{n-1} y = \dfrac{1}{(n-1)!} f^{(n-1)} + \dfrac{1}{(n-3)!} h_1 f^{(n-3)} + \dfrac{1}{(n-5)!} h_2 f^{(n-5)} + \cdots$

where the last terms are obtained to accord with the above conditions.

· Substituting the values $f^{(\varkappa)}$ we find the conditions to be

(n odd)
[35]
$$\begin{cases} H_{2\nu-2} + h_1 H_{2\nu-4} + h_2 H_{2\nu-6} + \cdots + h_{\nu-1} H_0 = 0 \\ (2\nu - 1) H_{2\nu-1} + (2\nu - 3) h_1 H_{2\nu-3} + (2\nu - 5) h_2 H_{2\nu-5} + \cdots \\ \qquad\qquad\qquad\qquad + h_{\nu-1} H_1 - h_\nu = 0 \end{cases}$$

(n even)
[36]
$$\begin{cases} H_{2\nu-1} + h_1 H_{2\nu-3} + h_2 H_{2\nu-5} + \cdots + h_{\nu-1} H_1 + h_\nu = 0 \\ 2\nu H_{2\nu} + (2\nu-2) h_1 H_{2\nu-2} + (2\nu-4) h_2 H_{2\nu-4} + \cdots + 2h_{\varkappa-1} H_2 = 0. \end{cases}$$

These conditions being satisfied $y = F(u)$ and we have two forms

$$D_u^2 F(u) - [n(n+1)pu + B] F(u) = 0$$

since finite for $u = ik' = \dfrac{\Omega'}{2}$.

A second solution being likewise obtained by making the substitution $n \sim -n$ the general integral may be written:

[37] · · · · · · · · $y = cF(u) + c'F(-u).$

Part III.

Integral as a Product.

Indirect Solution.

It will be shown in developing the forms for the case $n = 3$ that the original solution of M. Hermite as a sum will not be applicable in the forms given in the last chapter, when B is so taken as to give a value, v equal to zero, which leads to a second development in the form of a product, the eliments being as in the first case doubly periodic functions of the second species.

Assume that

$$[38] \quad \cdots \quad \cdots \quad y = \prod_{a=a\cdot b\cdots}^{n} \frac{\sigma(u+a)}{\sigma(u)\,\sigma(a)}\, e^{-u\zeta a},$$

where the product is composed of n factors obtained by taking a, b, c in place of a. The derivative of the logarithm is

$$\frac{y'}{y} = \sum \left[\zeta(u+a) - \zeta(u) - \zeta(a) \right] = \sum \frac{1}{2} \frac{p'u - p'a}{pu - pa}$$

while a second differentiation gives

$$\frac{y''}{y} - \left(\frac{y'}{y}\right)^2 = \sum \left[pu - p(u+a) \right].$$

From the first equation

$$\left[\frac{y'}{y}\right]^2 = \sum \frac{1}{4} \left(\frac{p'u - p'a}{pu - pa}\right)^2 + \frac{1}{2} \sum \frac{p'u - p'a}{pu - pa} \cdot \frac{p'u - p'b}{pu - pb}.$$

But the addition theorem gives:

$$\frac{1}{4} \frac{p'u - p'a}{pu - pa} = p(u+a) + pu + pa,$$

whence

$$\frac{y''}{y} = 2npu + \sum pa + \frac{1}{2} \sum \frac{p'u - p'a}{pu - pa} \cdot \frac{p'u - p'b}{pu - pb}.$$

In order to decompose the last term in this expression we write:

$$\frac{1}{2}\frac{p'u - p'a}{pu - pa}\cdot\frac{p'u - p'b}{pu - pb} = 2\,(pu + pa + pb)$$

$$+ \frac{p'a + p'b}{pa - pb}\,[\zeta(u + a) - \zeta(u + b) - \zeta u + \zeta b].$$

Take the value $u = (a + b)$, remembering the relations

$$p(-u) = + pu; \quad p'(-u) - p'(u); \quad \zeta(-u) = -\zeta(u).$$

Writing then $f(a + b)$ for the right hand member of the above equation under these conditions we get

$$f(a + b) = 2np\,(a + b) + \Sigma pa + 2p\,(a + b) + pa + pb$$

$$+ \frac{p'a + p'b}{pa - pb}\,[\zeta(-b) - \zeta(-a) - \zeta a + \zeta b]$$

$$= 2\,(n + 1)\,pu + 2\,\Sigma pa.$$

from which we see that in general we would have

$$\frac{y''}{y} = n\,(n + 1)\,pu + (2n - 1)\,\Sigma pa$$

the quantity in brackets being equal to zero.

If now we reunite the terms $\zeta(u + a) - \zeta a$, $\zeta(u + b) - \zeta b$ etc. in the general expression and make equal to zero the sum of their coefficients we obtain n equations of condition, namely, writing

$$pa = \alpha; \quad p'a = \alpha'; \quad pb = \beta; \quad p'b = \beta';$$

[39]

$$\frac{\alpha' + \beta'}{\alpha - \beta} + \frac{\alpha' + \gamma'}{\alpha - \gamma} + \frac{\alpha' + d'}{\alpha - d} + \cdots = 0$$

$$\frac{\beta' + \alpha'}{\beta - \alpha} + \frac{\beta' + \gamma'}{\beta - \gamma} + \frac{\beta' + d'}{\beta - d} + \cdots = 0$$

$$\frac{\gamma' + \alpha'}{\gamma - \alpha} + \frac{\gamma' + \beta'}{\gamma - \beta} + \frac{\gamma' + d}{\gamma - d} + \cdots = 0$$

$$- \quad - \quad - \quad - \quad - \quad - \quad -$$

If then we can solve the equations considered as simultanious

[40]

$$\alpha'^2 = 4\alpha^3 - g_2\alpha - g_3$$

$$\beta'^2 = 4\beta^3 - g_2\beta - g_3$$

together with the relation

$$(2n - 1)\,(\alpha + \beta + \gamma + \cdots) = B$$

we will satisfy the necessary conditions to enable us to write:

$$\frac{y''}{y} = n\,(n + 1)\,pu + B.$$

That is

$$y = \prod \frac{\sigma(u+a)}{\sigma(u)\,\sigma(a)}\, e^{-u\,\zeta\,a}$$

is a solution of Hermite's equation whatever be the value of B, provided $a, b, c \ldots$ fulfil the above conditions.

Solution for $n = 2$.

It is clear that, save for small values of n, an attempt to solve the above equations by the ordinary methods would give rise to insurmountable difficulties. The case $n = 2$ however, which is famous as affording a solution to the problem of a pendulum, constrained to move upon a sphere, can be readily solved as follows:
Given $n = 2$: we have the conditions

[41]
$$\alpha'^2 = p'^2 a = 4\alpha^3 - g_2\alpha - g_3$$
$$\beta'^2 = p'^2 b = 4\beta^3 - g_2\beta - g_3$$
$$pa + pb = \tfrac{1}{3} B \quad \text{or} \quad \alpha'^2 + \beta'^2 = 0.$$

Observe that by designating pb by $-\beta$ the above relations remain unaltered and that we may therefore write

$$4\alpha^3 - g_2\alpha - g_3 = -4\beta^3 + g_2\beta - g_3$$

or

$$4(\alpha^3 + \beta^3) - g_2(\alpha + \beta) = 0$$

whence

$$\alpha^2 - \alpha\beta + \beta^2 - \tfrac{1}{4} g_2 = 0.$$

But

$$\beta = \alpha - \tfrac{1}{3} B$$

whence the equation that determines the values of B.

[42] $\cdots$ $\tfrac{1}{9} B^2 - \tfrac{1}{3}\alpha B + \alpha^2 - \tfrac{1}{4} g_2 = 0$ and also $B = 0.$*)

If then $n = 2$ and a and b, the arguments of $\alpha = (-pa_2)$, are so taken that B shall have the values of the roots of equation (42) Hermite's equation will have the solution

---

*) If in this result we take $B = \tfrac{1}{3}\zeta$ we obtain the formula

$$\zeta^2 - \alpha\zeta + \alpha^2 - \tfrac{1}{4} g_2 = 0,$$

see Halphen II p. 131.

$$[43] \cdot \quad \cdot \quad \cdot \quad \cdot \quad \cdot \quad \cdot \quad y = c \, \frac{\sigma(u-a)\,\sigma(u+b)}{\sigma^2 u} \, e^{(\zeta a - \zeta b)u}$$

$$= c' \, \frac{d}{du} \left[\frac{\sigma(u-a+b)}{\sigma u} \, e^{(\zeta a - \zeta b)u} \right]$$

$$= \frac{d}{du} \left[\frac{\sigma(u+v)}{\sigma u} \, e^{(x - \zeta v)u} \right]$$

where $v = a + b$.*)

That our solution given above be complete we must obtain the corresponding values of x and u as follows:

$$y = \frac{d}{du} \left[\frac{\sigma(u+v)}{\sigma u} \, e^{(x - \zeta v)u} \right]$$

We have also

$$[44] \cdot \quad \cdot \quad \cdot \quad pv + pa + pb = \frac{1}{4} \frac{p'a + p'b}{pa - pb} = \left(\frac{p'a_2}{pa - pb} \right)$$

since

$$p'a = p'b = \alpha'.$$

Again we have

$$\zeta^2 - \alpha\zeta + \alpha^2 - \tfrac{1}{4} g_2 = 0$$

or

$$\zeta = \frac{\alpha}{2} \pm \frac{1}{2} \sqrt{g_2 - 3\alpha^2}.$$

Hence

$$p(a) = \frac{\alpha}{2} + \frac{1}{2} \sqrt{g_2 - 3\alpha^2}$$

$$p(b) = \frac{\alpha}{2} - \frac{1}{2} \sqrt{g_2 - 3\alpha^2}$$

whence

$$pa - pb = \sqrt{g_2 - 3\alpha^2}$$

$$pa + pb = \alpha$$

$$p'a_2 = -4\alpha^3 + g_2\alpha - g_3.$$

These values in (44) give:

$$[45] \cdot \quad \cdot \quad \cdot \quad \cdot \quad \cdot \quad \cdot \quad p(v) = \frac{-4\alpha^3 + g_2\alpha - g_3}{g_2 - 3\alpha^2} - \alpha$$

$$= \frac{\alpha^3 + g_3}{3\alpha^2 - g_2}.$$

*) The last is the form given for the expression $\cos CX + i \cos CY$ in the solution of the pendulum problem in the direct investigation of which one arrives at the expressions

$$\frac{d^2 X}{dt^2} = NX; \qquad \frac{d^2 Y}{dt^2} = NY; \qquad \frac{d^2 Z}{dt^2} = NZ + g$$

where N is found to be $3\tau^2(2pu - pa_2)$ which causes the solution to depend upon Lamé's functions.

If we take $\alpha = 2b$ we have

[46] $\cdot$ $\cdot$ $\cdot$ $\cdot$ $\cdot$ $\cdot$ $\cdot$ $\quad p(v) = \dfrac{8b^3 + g_3}{12b^2 - g_2} = \dfrac{b\varphi' - \varphi}{\varphi}$

where

$$\varphi = 4b^3 - g_2 b - g_3 \quad \text{and} \quad \varphi' = 12b^2 - g_2$$

For x we have:

[47] $\quad x = \zeta(a-b) + \zeta a - \zeta b = \dfrac{1}{2}\,\dfrac{p'(b-a) + p'b}{p(b-a) - pb}$

$$= \dfrac{1}{2}\,\dfrac{p'(b-a) - p'(a)}{p(b-a) - p(a)} = \dfrac{p'(b-a) + p'b - p'a}{2p(b-a) - pb - pa} = -\dfrac{p'v}{2pv - \alpha}$$

$$= \sqrt{\dfrac{3b\varphi' - \varphi}{\varphi'}}$$

since

$$p'a - p'b = 0$$
$$pa + pb = \alpha.$$

Combining these relations we obtain:

$$\dfrac{p'v}{2x} + pv = b$$

and

$$p'v = 2(b - pv)\sqrt{\dfrac{3b\varphi' - \varphi}{\varphi'}} = 2\left(b - \dfrac{b\varphi' - \varphi}{\varphi'}\right)\sqrt{\dfrac{3b\varphi' - \varphi}{\varphi'}}$$

$$= \dfrac{2\varphi}{\varphi'}\sqrt{\dfrac{3b\varphi' - \varphi}{\varphi'}}.^{*)}$$

Finally we observe that if $-u$ is substituted for u in Hermite's equation it remains unaltered which gives us the second solution, namely

[48] $\cdot$ $\cdot$ $\cdot$ $\cdot$ $\cdot$ $\cdot$ $\quad z = \prod \dfrac{\sigma(a-u)}{\sigma(a)\,\sigma(u)}\, e^{u\,\zeta a}$

x and v remaining as before.

Product of the Two Solutions.

It becomes evident from the illustration in the previous paragraph that while in general the theory involved in the solution just given holds it is practically inapplicable for other values of n than two or at most three whence one is led to a study of functions of the integral in the hope of discovering inherent properties

---

*) Compair results obtained by M. Halphen and obtained in a different manner, II p. 131 and 527.

that will lead to a more practical result. The first of such functions to command attention would be the product of the two integrals

[49] . . . . $Y = yz$

which we will proceed to develop as follows:

We would find from the integral z as in the case of y

$$\frac{z'}{z} = \sum [\zeta(a - u) - \zeta u + \zeta a]$$

and combining with

$$\frac{y'}{y} = \sum [\zeta(a + u) - \zeta(u) - \zeta a]$$

we obtain

$$\frac{y'}{y} - \frac{z'}{z} = \sum [\zeta(u + a) - \zeta(u - a) - 2\zeta a] = -\sum \frac{p'a}{pu - pa}$$

But

$$Y = yz = \prod \frac{\sigma(a + u)\,\sigma(a - u)}{\sigma^2 a\,\sigma^2 u} = \prod (pu - pa).^{*)}$$

Whence

$$yz' - zy' = \sum \frac{p'a}{pu - pa} \cdot yz$$

$$= \sum \frac{p'a}{pu - pa} \cdot \prod (pu - pa) = 2C$$

or

$$\sum \frac{p'a}{pu - pa} = \frac{2C}{\prod(pu - pa)},$$

C being a constant or expanding and writing $t = pu$ we have

[50] . . . $\dfrac{\alpha'}{t - \alpha} + \dfrac{\beta'}{t - \beta} + \dfrac{\gamma'}{t - \gamma} + = \dfrac{2C}{(t - \alpha)(t - \beta)(t - \gamma)\ldots}$

an identity independant of the value of t.

To determine $\alpha', \beta' \ldots$ multiply both members by $(t - \alpha)$, $(t - \beta)\ldots$ and take $t = \alpha, \beta \ldots$ for example

$$\alpha' + \frac{\beta'(t - \alpha)}{(t - \beta)} + \frac{\gamma'(t - \alpha)}{t - \gamma} + \cdots = \frac{2C}{(t - \beta)(t - \gamma)\ldots}$$

whence making $t = \alpha$ we have

[51] . . . . $\alpha' = \dfrac{2C}{(\alpha - \beta)(\alpha - \gamma)\ldots}$

and in a similar manner we find

$$\beta' = \frac{2C}{(\beta - \alpha)(\beta - \gamma)\ldots}$$

---

*) see theory of p and σ functions.

These values of α' and β'... determine the constants $a, b ...$ provided we can find the value of the constant C. It is also clear that C must be a constant involved in the relation

$$Y = y^2$$

and we are thus led first to a development of Y according to the powers of t and to the finding of the relation between the coefficients. Thus y becomes available in a practical form and C being determined as a function of Y and its derivatives we have our relation in a new form

[52] · · · · · · · · · · · $\quad y = \pm\sqrt{Y}.$

I expand these principles of M. Hermite*) (Annali di math.) and Halphen**) as follows:

Lamé's equation may be written

[53] · · · · · · · · · · $\quad y'' = Py$

where

$$P = n(n+1)\,pu + B \quad \text{and} \quad y = \sqrt{Y}.$$

Seeking the equation in terms of Y we write

$$Y' = 2yy'$$

whence

$$Y'' = 2y'^2 + 2yy'' = 2y'^2 + 2Py^2 = 2y'^2 + .2PY,$$

also

$$(y'' - 2PY)' = 4y'y'' = 4Pyy' = 2PY$$

whence

[54] · · · · · · · $\quad Y''' - 4PY' - 2P'Y = 0$

a linear differential equation in Y of the third order.

· From the theory of the linear differential equation, if y and z are solutions of (53) $\gamma y + qz$ will also be a solution γ and q being arbitrary constants, and we derive also as distinct solutions of the transformed (54) y^2, yz and z^2 obtained from the complex form $(\gamma y + qz)^2$

$$P' = n(n+1)\,p'u$$

and the transformed may be written:

[55] · · $\quad Y''' - 4[n(n+1)\,pu + B]\,Y' - 2n(n+1)\,p'uY = 0$

where

$$Y = \prod \frac{\sigma(a+u)\,\sigma(a-u)}{\sigma_2\,a\sigma^2 u} = \Pi\,(pu - ua).$$

This value indicates that (55) has n solutions in terms of $p(u)$

---

*) Bd. II. p. 498.

**) Bd. II. p. 498.

from which it follows also that Y may be written as an intire polynomial of the nth degree in $t = pu$. That is

[56] $\quad \cdot \quad \cdot \quad \cdot \quad Y = t^n + a_1 t^{n-1} + a_2 t^{n-2} + \cdots + a_{n-1} t + a_n.$

Equation [55] is written in terms of derivatives with respect to u whence to determine the coefficients in (56) we must express (55) also in terms of derivatives of $t = pu$ and equate the coefficients of like powers in the two identities thus obtained.

Take

$$\varphi = \varphi(t) = 4t^3 - g_2 t - g_3 = p'^2 u$$

whence

$$D_t u = \varphi^{-\frac{1}{2}}; \quad D_t^2 u = -\frac{1}{2} \varphi^{-\frac{3}{2}} \varphi'; \quad D_t^3 u = \frac{3}{4} \varphi^{-\frac{5}{2}} \varphi'^2 - \frac{1}{2} \varphi^{-\frac{3}{2}} \varphi''$$

and

$$D_u Y = D_t Y D_u t = \varphi^{\frac{1}{2}} D_t Y$$

$$D_u^2 Y = \frac{D_t u D_t^2 Y - D_t Y D_t^2 u}{(D_t u)^3}$$

$$D_u^3 Y = \frac{(D_t u)^2 D_t^3 Y - D_t u D_t^3 u D_t Y - 3 D_t u D_t^2 u D_t Y + 3 (D_t^2 u)^2 D_t Y}{(D_t u)^5}$$

$$= \varphi^{\frac{3}{2}} D_t^3 Y + \frac{3}{2} \varphi^{\frac{1}{2}} \varphi' D_t^2 Y - \frac{1}{2} \varphi^{\frac{1}{2}} \varphi'' D_t Y$$

These substitutions give:

[57] $(4t^3 - g_2 t - g_3) \dfrac{d^3 Y}{dt^3} + 3\left(6t^2 - \dfrac{1}{2} g_2\right) \dfrac{d^2 Y}{dt^2} - 4 \,|(n^2 + n - 3) t + B] \dfrac{dY}{dt}$
$$\qquad\qquad\qquad - 2n(n+1) Y = 0.$$

From [56] we obtain the values of these derivatives, namely

$$\frac{dY}{dt} = n t^{n-1} + a_1 (n-1) t^{n-2} + a_2 (n-2) t^{n-3} + a_3 (n-3) t^{n-4}$$
$$+ a_4 (n-4) t^{n-5} + \cdots$$

$$\frac{d^2 Y}{dt^2} = n(n-1) t^{n-2} + a_1 (n-1)(n-2) t^{n-3} + a_2 (n-2)(n-3) t^{n-4}$$
$$+ a_3 (n-3)(n-4) t^{n-5} + \cdots$$

$$\frac{d^3 Y}{dt^3} = n(n-1)(n-2) t^{n-3} + a_1 (n-1)(n-2)(n-3) t^{n-4}$$
$$+ a_2 (n-2)(n-3)(n-4) t^{n-5} + \cdots$$

and equating the coefficients to zero we have:

3*

$n - 3$: $4a_3(n-3)(n-4)(n-5) - g_2a_1(n-1)(n-2)(n-3)$

$- g_3n(n-1)(n-2) + 18a_3(n-3)(n-4)$

$- \frac{3}{2}g_2a_1(n-1)(n-2) - 4(n^2+n-3)(n-3)a_3$

$- 4Ba_2(n-2) - 2n(n+1)a_3 = 0$

$n - 4$: $4a(n-3)(n-4)(n-5) - g_2a_2(n-2)(n-3)(n-4)$

$- g_3a_1(n-1)(n-2)(n-3) + 18a_4(n-4)(n-5)$

$- \frac{3}{2}g_2a_2(n-2)(n-3) - 4(n^2+n-3)(n-4)a_4$

$- 4B(n-3)a_3 - 2n(n+1)a_4 = 0$

- - - - - - - - - - - - - - - - - -

$n - k$: $4a_k(n-k)(n-k-1)(n-k-2)$

$- g_2a_{k-2}(n-k+2)(n-k+1)(n-k)$

$- g_3a_{k-3}(n-k+3)(n-k+2)(n-k+1)$

$+ 18a_k(n-k)(n-k-1)$

$- \frac{3}{2}g_2a_{k-2}(n-k+2)(n-k+1)$

$- 4(n^2+n-3)(n-k)a_k - 4B(n-k+1)a_{k-1}$

$- 2n(n+1)a_k = 0.$

From the last value we pass to the nth by writing

$$k = n - \mu$$

whence the recurring formula:

[58] $2(n-\mu)(2\mu+1)(\mu+n+1)a_{n-\mu} + 4(\mu+1)Ba_{n-\mu-1}$

$+ \frac{1}{2}g_2(n+1)(\mu+2)(2\mu+3)a_{n-\mu-2}$

$+ g_3(\mu+1)(\mu+2)(\mu+3)a_{n-\mu-3} = 0$

from which equation we find the unknown coefficients a_i by making

$$\mu = n-1, \quad n-2, \ldots \quad \text{or} \quad k = 1.2\ldots$$

These results are simplified by employing the notation introduced by Brioschi, namely:

$$S = t-b: \quad b = \frac{1}{n(2n-1)}B. \quad \varphi(t) = 4t^3 - g_2t - g_3 = \varphi. \quad t = pu,$$

by means of which the above forms are expressed as follows:

[59] $\left[4S^3 + \frac{1}{2}\varphi''S^2 + \varphi'S + \varphi\right]\dfrac{d^3Y}{dS^3} + \left(18S^2 + \frac{3}{2}\varphi''S + \frac{3}{2}\varphi'\right)\dfrac{d^2Y}{dS^2}$

$-\left[4(n^2+n-3)S + \dfrac{n^2-1}{2}\varphi''\right]\dfrac{dY}{dS} - 2n(n+1)Y = 0$

[60] $\cdots \cdots \quad Y = S^n + A_2S^{n-2} + A_3S^{n-3} + \cdots + A_n$

[61] $2 (n - \mu) (2\mu + 1) (\mu + n + 1) A_{n-\mu}$

$$= 12 (\mu + 1) (\mu + 1 - n) (\mu + 1 + n) b A_{n-\mu-1}$$

$$+ \tfrac{1}{2} (\mu + 1) (\mu + 2) (2\mu + 3) \varphi' (b) A_{n-\mu-2}$$

$$+ (\mu + 1) (\mu + 2) (\mu + 3) \varphi (b) A_{n-\mu-3}.$$

Taking $\mu = n - 1$ we find $A_1 = 0$

$\mu = n - 2:$ $A_2 = \dfrac{n (n - 1)}{8 (2n - 3)} \varphi' (b)$

$\mu = n - 3:$ $A_3 = \dfrac{n (n - 2)}{12 (2n - 5)} \varphi (b) - \dfrac{n (n - 1) (n - 2)}{2 (2n - 3) (2n - 5)} b \varphi' (b).$

And the term containing the highest power of B is obtained as follows:

$\mu = n - 2:$ $2 \cdot 2 (2n - 3) (2n - 1) a_2 = - 4 (n - 1) B$

or $a_2 = - \dfrac{(n - 1) B}{(2n - 1) (2n - 3)}$

$\mu = n - 3:$ $a_3 = \dfrac{(n - 2) B^2}{3 (2n - 1) (2n - 5)}$

$\mu = n - 4:$ $a_4 = \dfrac{(n - 2) (n - 3) B^3}{2 \cdot 3 \cdot (2n - 1) (2n - 3) (2n - 5) (2n - 7)} + \cdots$

$\mu = n - 5:$ $a_5 = - \dfrac{(n - 3) (n - 4) B^4}{2 \cdot 3 \cdot 5 (2n-1)(2n-3)(2n-5)(2n-7)(2n-9)} + \cdots$

[62] $\mu = 1:$ $a_{n-1} = \dfrac{(- 1)^n B^n}{[3 \cdot 5 \cdot 7 \cdots 2n - 1]^2} + \cdots$

Direct Solution.

Having $Y = yz$, we are enabled to obtain a rigid and direct solution of **Hermite's** equation in the form of a product as follows:

In addition to Y we have:

$$Y' = yz' + zy' \quad \text{and} \quad yz' - zy' = 2C.$$

whence

$$2 yz' = 2C + Y', \quad \text{or} \quad \frac{z'}{z} = \frac{2C + Y'}{2 Y}$$

and

$$- 2 zy' = 2C - Y', \quad \text{or} \quad \frac{y'}{y} = \frac{Y' - 2C}{2 Y}.$$

whence

$$\frac{yy'' - y'^2}{y^2} = \frac{y''}{y} - \left(\frac{y'}{y}\right)^2 = \frac{YY'' - Y'^2}{2 Y^2}$$

or

$$\frac{y''}{y} = \frac{2 YY'' - Y'^2 + 4C^2}{4 Y^2}.$$

This value in Hermite's equation gives:

[63] $\cdots\quad\cdot\ 2\,YY'' - Y'^2 + 4C^2 = [n(n+1)pu + B]\,4\,Y^2.$

Whence we derive the value of C sought, namely

[64] $\cdots\quad\cdot\ 4C^2 = Y'^2 - 2\,YY'' + 4[n(n+1)pu + B]\,Y^2.$

Let $\alpha, \beta, \gamma \cdots = pa, pb, py \cdots$ be roots of Y.

Then

$$Y_{u=a\cdot b\cdots} = t^n + a_1 t^{n-1} + \cdots = 0$$

$$Y'_{u=a\cdot b\cdots} = n t^{n-1}t' + a_1(n-1)t^{n-2}t' + \cdots = 0$$

or

$$\frac{dY}{du} = np(u)^{n-1}p'u.$$

Whence

$$Y'_u = p'u\frac{dY}{dt}$$

and

$$4C^2 = p'^2(a)\left[\frac{dY}{dt}\right]_{t=\alpha}^2 = Y'_\alpha = p'^2(b)\left(\frac{dY}{dt}\right)_{t=\beta}^2 = Y'_\beta \cdots$$

But from algebra we have

$$\left[\frac{dY}{dt}\right]_{t=\alpha} = (\alpha - \beta)(\alpha - \gamma)\cdots$$

Whence

[65] $\cdots\quad\cdot\quad\cdot\ 2C = \alpha'(\alpha - \beta)(\alpha - \gamma)\cdots$

with like expressions for the other roots which we observe are the values obtain before (see [51]), namely

$$\alpha' = \frac{2C}{(\alpha - \beta)(\alpha - \gamma)\cdots}$$

$$\beta' = \frac{2C}{(\beta - \alpha)(\beta - \gamma)\cdots}$$

To obtain Y we have:

$$2C = Y'_a = Y'_b = Y'_c = \cdots$$

$\pm\,a,\ \pm\,b,\ \pm\,c$ being the roots of $Y = f(u)$.

We have also:

$$2C = yz' - zy'$$

$$= \sum[\zeta(u+a) - \zeta(u-a) - 2\zeta(a)]yz$$

or

$$\frac{2C}{Y} = \sum[\zeta(u+a) - \zeta(u-a) - 2\zeta(a)].$$

But

$$\frac{p'u}{2(pu - pa)} = \tfrac{1}{2}[\zeta(u+a) + \zeta(u-a) - 2\zeta a]$$

whence

$$-\frac{C}{Y} = \sum\left[\zeta(u+a) - \frac{p'u}{2(pu-pa)} - \zeta u - \zeta a\right]$$

$$= \frac{d}{du}\sum[\log\sigma(u+a) - \log\sqrt{pu-pa} - \log\sigma u - u\zeta a]$$

$$= \frac{d}{du}\log\prod\frac{\sigma(u+a)}{\sqrt{pu-pa}\cdot\sigma u}\,e^{-u\zeta a}$$

$$= \frac{d}{du}\log\prod\frac{\sigma(u+a)}{\sigma u}\,e^{-u\zeta a} - \frac{d}{du}\log\prod\sqrt{pu-pa}.$$

But

$$\prod\sqrt{pu-pa} = y^{\frac{1}{2}}z^{\frac{1}{2}}$$

$$\therefore -\frac{C}{Y} = \frac{d}{du}\log\prod\frac{\sigma(u+a)}{\sigma u}\,e^{-u\zeta a} - \frac{d}{du}\log y^{\frac{1}{2}}z^{\frac{1}{2}}$$

$$= \frac{d}{du}\log\prod\frac{\sigma(u+a)}{\sigma u}\,e^{-u\zeta a} - \frac{1}{2}\frac{yz'+zy'}{yz}$$

$$= -\frac{Y'}{2Y} + \frac{d}{du}\log\prod\frac{\sigma(u+a)}{\sigma(u)}\,e^{-u\zeta a}.$$

Whence

$$\frac{Y'-2C}{2Y} = \frac{y'}{y} = \frac{d}{du}\log\prod\frac{\sigma(u+a)}{\sigma u}\,e^{-u\zeta a}$$

or

$$\log y = \log\prod\frac{\sigma(u+a)}{\sigma u}\,e^{-u\zeta a} - \log C$$

$$C = \Pi\sigma a.$$

Whence the value of y is obtained directly, namely

[66] $\qquad\qquad y = \prod\frac{\sigma(u+a)}{\sigma u\,\sigma a}\,e^{-u\zeta a}.$

The third method of integration is then the following:

Calculate the polynomial Y by the aid of the relation [58] or [61] from which derive the Constant C^2 by means of equation [64] extracting the square root to obtain C and finally obtain the constants

$$p'a = \frac{2C}{(\alpha-\beta)(\alpha-\gamma)\cdots}, \qquad p'b = \frac{2C}{(\beta-\alpha)\beta-\gamma)\cdots}$$

when $\alpha = \gamma a$, $B = \gamma b \ldots$ are the roots of Y.

These relations determine the arguments $a \cdot b \cdot c \ldots$, having which the solution is

$$y = \prod\frac{\sigma(u+a)}{\sigma u}\,e^{-u\zeta a}.$$

If we take the second root of C^2 we obtain the integral z obtained also from y by changing u into $-u$.

Determination of Y for $n = 3$.

The foregoing solution while complete and rigid from a theoretical standpoint needs to be greatly perfected before it becomes practically applicable. It is indeed but another example, the invariant theory being a second of the fact that it is often an easier task to obtain a general than an explicit form. Having determined the explicit forms for $n = 2$ let us attempt to apply the above rule to the next case $n = 3$.

From (60) and (61) we obtain.

Given $n = 3$

$$Y_{n=3} = S^3 + A_2 S + A_3$$

where

$$A_2 = \frac{n(n-1)}{8(2n-3)} \, \varphi'(b) = \tfrac{1}{4}\, \varphi'(b) = \tfrac{1}{4}\left(12b^2 - \tfrac{1}{4} g_2\right)$$

$$A_3 = \frac{n(n-2)}{12(2n-5)} \, \varphi(b) - \frac{n(n-1)(n-2)}{2(2n-3)(2n-5)} \, b\varphi'b = \tfrac{1}{4}\, \varphi(b) - b\varphi'b$$

$$= -\tfrac{1}{4}\left(44b^3 - 3g_2 b + g_3\right).$$

Again $S = t - b$

$$\therefore \varphi(t) = 4(S + b)^3 - g_2(S + b) - g_3$$
$$= 4S^3 + 12bS^2 + 12b^2S + 4b^3 - g^2 S - bg_2 - g_3$$
$$= 4S^3 + 12bS^2 + (12b^2 - g_2) S + 4b^3 - bg_2 - g_3$$
$$= 4S^3 + 12bS^2 + \varphi'S + \varphi.$$

$$\therefore S^3 = \tfrac{1}{4}\, \varphi(t) - 3bS^2 - \tfrac{1}{4}\, \varphi'S - \tfrac{1}{4}\, \varphi.$$

Hence

[67] $\cdot \ \cdot \ Y_{n=3} = S^3 + A_2 S + A_3$
$$= S^3 + \tfrac{1}{4}\varphi'S + \tfrac{1}{4}\varphi - b\varphi'$$
$$= S^3 + \left(3b^2 - \tfrac{1}{4}g_2\right) S - \tfrac{1}{4}\left(44b^3 - 3g_2 b + g_3\right)$$
$$= \tfrac{1}{4}\varphi(t) - b\left(\varphi' + 3S^2\right)$$
$$= \tfrac{1}{4}\varphi(t) - b[\varphi' + 3(t - b)^2]$$
$$= t^3 - 3bt^2 + \left(6b^2 - \tfrac{1}{4}g_2\right)t - \left(15b^3 - g_2 b + \tfrac{1}{4}g_3\right).$$

Whence

$$Y' = 3S^2 + A_2, \quad Y'' = 6S, \quad 2YY'' = 12S(S^3 + A_2 S + A_3)$$

and substituting in (64) we have

[68] $\cdot \ \cdot \ \cdot \ \cdot \ C^2 = \tfrac{1}{4}(3S^2 + A_2)^2 - 3S(S^3 + A_2 S + A_3)$
$$\qquad\qquad + 3(4S + qb)(S^3 + A_2 S + A_3)^2.$$

To attempt to extract the square roots of this equation in accordance with the theory, C^2 being expressed as an equation of the 7th degree in S or t were clearly impossible without some further knowledge of the properties of C. To arrive at such knowledge we are led ultimately back to a study of the special functions of Lamé.

Part IV.

The Special Functions of Lamé.

Functions of the First Sort.

Lamé derived originally functions of three different sorts, values for y, depending on the value of n and corresponding in each case to a specific value of B, the chief peculiarity being that for these values y is doubly periodic.

The functions of the first class are characterized as developable in the form

[69] $\quad \cdot \; \cdot \; \cdot \; \cdot \quad y = p^{(n-2)} + a_1 p^{(n-4)} + a_2 p^{(n-6)} + \cdots$

and that such an integral may exist is seen from the following:

Writing the corresponding function of the same sort $y \cdot p(u)$ we have

$$n(n+1)yp(u) = p^{(n)} + A_1 p^{(n-2)} + A_2 p^{(n-4)} + \cdots$$

whence by subtraction

$$y'' - n(n+1)yp = (a_1 - A_1)p^{(n-2)} + (a_2 - A_2)p^{(n-4)} + \cdots$$
$$= By$$

that is a function of the first sort will be a root of Hermite's equation provided

$$a_1 - A_1 = B : a_2 - A_2 = Ba_1 : a_3 - A_3 = Ba_2 \quad \text{etc.}$$

Where the quantities (A) are linear functions of the quantities (a). But since the number of these condition equations is greater by unity than the number of unknown (a) it follows that upon their ellimination we obtain an equation in B whose degree will equal the number of equations, that is $\frac{1}{2}n + 1$ if n is even and $\frac{1}{2}(n-1)$ if n is uneven:

For example take $n = 2$, whence $y = p + a_1$ and $y'' = p''$ and we derive

$$p'' - 6(p + a_1)p = Bp - Ba_1$$

or

$$Ba_1 + \tfrac{1}{2}g_2 = 0, \quad \text{also} \quad 6a_1 + B = 0$$

whence

$$a_1 = -\tfrac{1}{6} B$$

and we find

$$y = p - \tfrac{1}{6} B \qquad \text{where} \qquad B^2 - 3g_2 = 0.$$

Again let $n = 3$ in which case the equation in B would be of degree $\tfrac{1}{2}(n - 1) = 1$, that is $B = 0$, for which value we have at once

$$y = p'(u).$$

Substituting indeed this value in Hermite's equation for $n = 3$ we derive at once

$$p''' - 12p'p = 0$$

a well known identity.

Define $(P = 0)$ equal to the equation in B of degree $\tfrac{1}{2}(n - 1)$ that in any case determines the values of B giving rise to an integral of the first sort.

We have then that when $P = 0$ the general solution of Hermite as a sum has in place of $f(u)$ the $p(u)$ and may be written

$$[70] \quad \cdots \quad (-1)^n y = \frac{1}{(n-1)!} p^{(n-2)}(u) + \frac{1}{(n-3)!} h_1 p^{(n-4)}(u)$$
$$+ \frac{1}{(n-5)!} h_2 p^{(2-6)}(u) + \cdots$$

the coefficients being the same as in the corresponding general development.[*])

Functions of the Second Sort.

To attain a function of the second sort assume that n is odd and that the solution has the form

$$[71] \quad \cdots \quad \cdots \quad \cdots \quad y = z\sqrt{pu - e_\alpha} \qquad \qquad \alpha = 1. 2. 3$$

where z may be developed in the form

$$z = p^{(n-3)} + a_1 p^{(n-5)} + a_2 p^{(n-7)}$$

an equation in p differing from (70) in the degree of the derivatives only. Proceeding as in the former case by substituting in Hermite's equation one finds that the solution holds provided B be now taken equal to any one of the roots of a perfectly determined equation of degree $\tfrac{1}{2}(n + 1)$, the right hand member of which we will define as Q_α which is equal to zero.

---

[*]) see (34) and (26).

Writing for convenience Hermite's equation in terms of the derivatives of z with respect to pu by aid of the identity

$$p'^2 = 4p^3 - g_2 p - g_3$$

we have*)

[72] $\cdot$ $\cdot$ $(4p^3 - g_2 p - g_3) \dfrac{d^2 z}{dp^2} + \left(10 p^2 + 4 e_\alpha p + 4 e_\alpha^2 - \dfrac{3}{2} g_2\right) \dfrac{dz}{dp}$

$$= [(n-1)(n+2)p + B - e_\alpha] z.$$

Take now for example $n = 3$. whence

$$z = p + a \qquad \frac{dz}{dp} = 1 \qquad \frac{d^2 z}{dp^2} = 0$$

and (72) becomes

$$10 p^2 + 4 e_\alpha p + 4 e_\alpha^2 - \frac{3}{2} g_2 = (10 p + B - e_\alpha)(p + a_1)$$

and differentiating we have

$$4 e_\alpha = 10 a_1 + B - e_\alpha$$

or

$$a_1 = \frac{1}{2} e_\alpha - \frac{1}{10} B$$

whence

$$z = p + \frac{1}{2} e_\alpha - \frac{1}{10} B$$

and

[73] $\cdot$ $\cdot$ $\cdot$ $\cdot$ $\cdot$ $Q_\alpha = B^2 - 6 e_\alpha B + 45 e_\alpha^2 - 15 g_2 = 0$

an equation whose degree is

$$\frac{1}{2}(n+1) = 2,$$

and as α may have the values 1, 2 or 3 we have in all six values of B giving a doubly periodic solution of the second sort and determined by an equation of the sixth degree defined as

[74] $\cdot$ $\cdot$ $\cdot$ $\cdot$ $\cdot$ $\cdot$ $\cdot$ $\cdot$ $\cdot$ $\cdot$ $Q = Q_1 Q_2 Q_3.$ $\cdot$ $\cdot$

Functions of the Third Sort.

We have finally solutions that are doubly periodic of a third sort the integral being written in the form:

[75] $\cdot$ $\cdot$ $\cdot$ $\cdot$ $\cdot$ $\cdot$ $\cdot$ $y = z \sqrt{(pu - e_\beta)(pu - e_\alpha)}$

where n is restricted to an even member and z has the form

$$z = p^{(n-4)} + a_1 p^{(n-6)} + a_2 p^{(n-8)} + \cdots$$

and a similar analysis to the former cases shows that this solution holds when B is the root of a determinate equation whose degree is $\frac{1}{2} n$.

---

*) compair transformation p. 35.

Part V.

Reduction of the Forms when n equals three.

Identity of Solutions.

Having developed in the foregoing the necessary underlying principles we return to the case where n equals three, that is to a determination of the integral of the equation

[76] $\cdots \cdots \cdots \cdots \quad y'' = [12p(u) + B]y$

where B is to be arbitrarily chosen.

The first form obtain from (32) is

$$y = \tfrac{1}{2}f'' + h_1 f$$

and from the first of equations (26) we have

$$h_1 = -\frac{B}{10} \cdot \text{ where } \quad B = 15b$$

Hence disregarding the constant the integral is

[77] $\cdots \cdots \cdots \cdots \quad y = f'' - 3bf$

where

$$f = \frac{\sigma(u + v)}{\sigma(u)\,\sigma(v)}\, e^{(x - \zeta v)u}$$

and x and v satisfy the conditions (35)

[78] $\cdots \cdots \cdots \quad \begin{cases} H_2 + h_1 H_0 = 0 \\ 3H_3 + h_1 H_1 = h_2 \end{cases}$

where

$$x = \xi v - \xi a - \xi b - \xi c$$
$$v = a + b + c.$$

$\left.\begin{array}{c}\\ \\ \\ \\ \\ \\ \\ \\ \\ \\ \end{array}\right\}$ 1st Solution.

(p. 17 and p. 16.)

The second form .obtained from (66) is

[79] · · · $y = \prod \dfrac{\sigma(u + a)}{\sigma a \, \sigma u} \, \varepsilon^{- u \zeta a} = \prod \dfrac{\sigma(u - a)}{\sigma(a) \, \sigma(u)} \, \varepsilon^{u \zeta a}$

$= \dfrac{\sigma(u - a) \, \sigma(u - b) \, \sigma(u - c)}{\sigma(a) \, \sigma(b) \, \sigma(c) \, \sigma^3 \omega} \, \varepsilon^{(\zeta a + \zeta b + \zeta c) u}$

where

$$\alpha' = p'(a) = \frac{2C}{(\alpha - \beta) \, (\alpha - \gamma)} \quad ,$$

$$\bullet \beta' = p'(b) = \frac{2C}{(\beta - \alpha) \, (\beta - \gamma)}$$

$$\gamma' = p'(c) = \frac{2C}{(\gamma - \alpha) \, (\gamma - \beta)}$$

and

$$C = \pm \sqrt{Y}; \quad Y = S^3 + A_2 S + A_3 \qquad \text{(p. 40.)}$$

$$S = t - b; \quad A_2 = \tfrac{1}{4}\left(12 b^2 - \tfrac{1}{4} g_2\right); \quad A_3 = - \tfrac{1}{4}(44 b^3 - 3 g_2 b + g_3).$$

The transformation of form (79) to form (77) may be accomplished as follows. Taking the eliments we have

$$\frac{\sigma(u + a)}{\sigma u \, \sigma a} e^{- u \zeta a} = \frac{1}{u} - \frac{u}{2} \, p a + \cdots$$

$$\frac{\sigma(u + b)}{\sigma(u) \, \sigma b} e^{- u \zeta b} = \frac{1}{u} - \frac{u}{2} \, p b + \cdots$$

$$\frac{\sigma(u + c)}{\sigma u \, \sigma c} e^{- u \zeta c} = \frac{1}{u} - \frac{u}{2} \, p c + \cdots$$

whence

$$y = \frac{\sigma(u + a) \, \sigma(u + b) \, \sigma(u + c)}{\sigma(a) \, \sigma(b) \, \sigma(c) \, \sigma^3 u} \, e^{-(\zeta a + \zeta b + \zeta c) u}$$

$$= \left[\frac{1}{u^2} - \frac{1}{2}(p a + p b) + \right]\left(\frac{1}{u} - \frac{u}{2} p(c) + \right)$$

Take

$$f = \frac{\sigma(u + a + b + c)}{\sigma(a + b + c) \, \sigma u} e^{- u(\zeta a + \zeta b + \zeta c)} = \frac{\sigma(u + v)}{\sigma u \, \sigma v} e^{(x - \zeta v) u}$$

$$= \frac{1}{u} - \frac{u}{2}(p a + p b + p c) + \cdots$$

$$f' = - \frac{1}{u^2} - \frac{1}{2}(p a + p b + p c) + \cdots$$

$$f'' = \frac{2}{u^3} + \cdots$$

Whence we observe that we may write

$$y = \tfrac{1}{2}[f'' u - (p a + p b + p c) f u].$$

But

$$p a + p b + p c = \tfrac{1}{5} B = 3 b$$

and, disregarding the factor $\frac{1}{2}$, we obtain the first form:

$$y = f'' - 3bf.$$

Having then a method of reduction the determination of $a : b : c$ is involved in the determinate of ν.

Determination of x and ν. First Method.

To this end we have from (31) and (26)

$$H_0 = x; \quad H_1 = \tfrac{1}{2}(x^2 + P_2); \quad H_2 = \tfrac{1}{6}(x^3 + 3 P_2 x + P_3)$$

and also

$$\begin{cases} h_1 = -\dfrac{B}{10} \\[2mm] h_2 = \dfrac{B^2}{120} - \dfrac{g_2}{20} \end{cases}$$

whence relations (78) become

$$\tfrac{1}{3}(x^3 + 3 P_2 x + P_3) - \frac{B}{5} x = 0$$

$$\tfrac{1}{2}(x^4 + 6 P_2 x^2 + 4 P_3 x + P_4) - \frac{B}{5}(x^2 + P_2) = \frac{B^2}{30} - \frac{g_2}{5}$$

set

$$l = \tfrac{1}{5}.B \quad \text{or} \quad h_1 = -\frac{l}{2}$$

and take from (p. 24)

$$P_2 = p\nu; \quad P_3 = -p\nu; \quad P_4 = -3p^2\nu + \tfrac{3}{5}g_2$$

Which values reduce our relations to the form

[80]
$$\begin{array}{l} \text{(a)} \left| x^3 - 3p(\nu)x - p'(\nu) - 3lx = 0 \right. \\ \text{(b)} \left| x^4 - 6p(\nu)x^2 - 4p'(\nu)x - 3p^2(\nu) - 2l + 2lp(\nu) = \dfrac{5 l^2}{3} - g_2 \right. \end{array}$$

which are reduced forms of the equations of condition that $y = F_1(x)$ be a solution in addition to which we have the identity

$$p'(\nu)^2 = 4p^3(\nu) - g_2 p(\nu) - g_3$$

and the useful relation

$$H_1 = \tfrac{1}{2}(x^2 - p(\nu)), \quad \text{or} \quad p(\nu) = x^2 - 2H_1.$$

The product of equations (80) is an equation of the seventh degree in x the roots of which are functions of ν and B and hence the values of B that will reduce x to zero are in number not more than seven.

But when x equals zero (and $\nu = w_\lambda$), y is in general a doubly periodic function and the doubly periodic special functions of Lamé are in all seven in number for n equals three one being of the first sort and six of the second.

It follows then that by elliminating $p(\nu)$ and $p'(\nu)$, we should obtain x as a function of Φ where Φ is a function of B the vanishing of which will be the condition for the special functions of Lamé.

This complicated ellimination, suggesting the practical uselessness of this method for any higher value of n is performed as follows.

Multiplying the first equation by four and subtracting we obtain

$$3x^4 - 6p(\nu)x^2 - 10lx^2 - 2lp(\nu) + 3p^2(\nu) = -\frac{5l^2}{3} + g_2$$

whence the relation

$$p(\nu) = x^2 - 2H_1$$

gives

(c) $\qquad 36H_1^2 - 36bx^2 + 12lH_1 + 5l^2 - 3g_2 = 0.$

Again from (b) and the identity

$$p'(\nu)^2 = (3bx + 3p(\nu)x - x^3)^2 = 9b^2x^2 + 9p^2(\nu)x^2 + x^6 + 18bp(\nu)x^2$$
$$- 6bx^4 - 6p(\nu)x^4$$

$$= 9b^2x^2 + 9x^2(x^4 - 4x^2H_1 + 4H_1^2) + x^6 + 18bx^2(x^2 - 2H_1)$$
$$- 6bx^4 - 6x^4(x^2 - 2H_1)$$

$$= 4(x^6 - 6x^4H_1 + 12x^2H_1^2 - 8H_1^3) - g_2(x^2 - 2H_1) - g_3$$

or multiplying by 9

(d) $\qquad 81l^2x^2 - 108x^2H_1^2 + 108lx^4 - 9 \cdot 36lH_1x^2 + 9 \cdot 32H_1^3$
$$+ 9g_2x^2 - 18g_2H_1 + 9g_3 = 0.$$

From (a), (b) and the value for $p(\nu)$

$$x^4 - 6x^2(x^2 - 2H_1) - 4x(x^3 - 3p(\nu)x - 3bx) - 3(x^4 - 4x^2H_1 + 4H_1^2)$$
$$- 2lx^2 + 2l(x^2 - 2H_1) = \frac{3l^2}{3} - g_2$$

or

(e) $\qquad 12lx^2 - 12H_1 - 4bH_1 = \frac{5l^2}{3} - g_2$

and multiplying (e) by 3 and $8H_1$ it becomes

(f) $\qquad 36 \cdot 8lx^2H_1 - 36 \cdot 8H_1^3 - 96lH_1^2 = 40l^2H_1 - 24g_2H_1$

whence from (c) elliminating H_1^3

(g) $\qquad 81l^2x^2 - 108x^2H_1^2 + 108lx^4 - 36lH_1x^2 - 96lH_1^2 = 40l^2H_1$
$$- 6g_2H_1 - 9g_2x^2 - 9g_3.$$

Whence a further combination with (c) gives

(h) $72 l^2 x^2 - 72 l H_1^2 - 32 l^2 H_1 + 6 g_2 H_1 + \dfrac{10 l^3}{3} + 9 g_3 - 2 l g_2 = 0$

and again

(i) $8 l^2 H_1 - 6 g_2 H_1 - \dfrac{40 l^3}{3} - 9 g_3 + 8 l g_2 = 0.$

Whence

[81] . . .   . . . $H_1 = \dfrac{10 l^3 - 6 l g_2 + \dfrac{2 l}{4} g_3}{6 \left(l^2 - \dfrac{3}{4} g_2 \right)}$

$$= \frac{10 l^3 - 8 a_1 l - b_1}{6 (l^2 - a_1)}$$

where

$$a_1 = \frac{3 g_2}{4} \quad \text{and} \quad b_1 = -\frac{2 l}{4} g_3.$$

From this value of H_1 we have by substituting in (c)

[82] . . $x^2 = \dfrac{125 l^6 - 210 a_1 l^4 - 22 b_1 l^3 + 93 a_1^2 l^2 + 18 a_1 b_1 l + b_1^2 - 4 a_1^3}{36 l (l^2 - a_1)^2}$

$$= \frac{4 (l^2 - a_1)^3 + (11 l^3 - 9 a_1 l - b_1)^2}{36 l (l^2 - a_1)^2}$$

$$= \frac{\Phi(l)}{S D^2}$$

where

$$\Phi(l) = 125 l^6 - 210 a_1 l^4 - 22 b_1 l^3 + 93 a_1^2 l^2 + 18 a b_1 l + b_1^2 - 4 a_1^3$$

$$S = 36 l, \quad D = (l^2 - a_1), \quad l = \tfrac{1}{5} B = 3 b$$

$$a_1 = \tfrac{3 g_2}{4} = \tfrac{1}{\lambda^2}(1 - k^2 + k^4); b_1 = -\tfrac{2 l}{4} g_3 = \tfrac{1}{\lambda^3}(1 + k^2)(2 - k^2)(1 - 2 k^2).^*)$$

$\Phi(l) = 0$ is then the condition for the existence of the special functions of Lamé the seventh value of B, as we have already seen (p. 43), being $B = 0$.

$\Phi(l)$ must then be $Q(l)$ times a constant and as we have seen that Q is separable into three factors of the second degree it follows that $\Phi(l)$ is a reducable equation of the sixth degree.[**])

Moreover if we make the transformation

$$l = l_1^{\tfrac{1}{3}} \xi$$

*) The expressions used here are essentially the same as those of M. Hermite in his celebrated Memoir. The following reduction of the function $\varphi(l)$ is also indicated by Hermite.

**) It is interesting to note that it is not given under the head of reducable forms of the sixth degree by either Clebsch or Gordan.

4

the coefficients of Φ all reduce to functions of the absolute invariant of the fourth degree

$$c = \frac{a_1}{b_1^{\frac{2}{3}}} \quad \text{and} \quad c^3 = \frac{a_1^3}{b_1^2} = \frac{1}{108}\frac{g_2^3}{g_3^2} = \frac{(1 - k^2 + k^4)^3}{(1 + k^2)^2 (2 - k^2)^2 (1 - 2k^2)^2}$$

and we have the form:

[83] $\quad\cdot\quad \Phi(\xi_1) = \frac{\Phi(b^{\frac{1}{3}}\xi)}{b^2} = 125\xi^6 - 210c\xi^4 - 22\xi^3 + 93c^2\xi^2 + 18c\xi$
$$+ 1 - 4c^3 = 0.$$

If then this equation be written in its expanded form in terms of the modulus k it will not be difficult to see by inspection (for rigorous proof see p. 56) that if we write

[84] $\quad\cdot\quad\cdot\quad\cdot\quad\cdot\quad\cdot\quad\cdot\quad\cdot\quad\cdot\quad \Phi = \Phi_1\Phi_2\Phi_3$

these factors of Φ corresponding to the special functions of the second sort are, as given by M. Hermite:

[85] $\quad\cdot$
$$\begin{aligned}
\Phi_1 &= 5l^2 - 2(k^2 - 2)l - 3k^4 \\
\Phi_2 &= 5l^2 - 2(1 - 2k^2)l - 3 \\
\Phi_3 &= 5l^2 - 2(1 + k^2)l - 3(1 - k^2)^2.
\end{aligned}$$

When $\Phi = 0$ we have $x = 0$ whence, as before stated, $\Phi = 0$ is a necessary condition for the existence of a doubly periodic function. But in order to be a sufficient condition it must involve a definite value of ν, that is ν must be a half-period. That this is the case, although the reverse as we shall find later does not hold, is seen by a determination of ν as follows:

We have (p. 47)

$$\begin{aligned}
p(\nu) &= x^2 - 2H_1 \\
&= \frac{\Phi(l) - 12l(l^2 - a_1)(10l^3 - 8a_1 l - b_1)}{36l(l^2 - a_1)^2}.
\end{aligned}$$

Define

$$\begin{aligned}
\psi(l) &= \Phi(l) - 12l(l^2 - a_1)(10l^3 - 8a_1 l - b_1) \\
&= 5l^6 + 6a_1 l - 10b_1 l^3 - 3a_1^2 l^2 + 6a_1 b_1 l + b_1^2 - 4a_1^3.
\end{aligned}$$

Whence we write

[86] $\quad\cdot\quad\cdot\quad\cdot\quad\cdot\quad p(\nu) = k^2\operatorname{sn}^2\omega - \frac{1 + k^2}{3} = \frac{\psi(l)}{36l(l^2 - a_1)^2}.$

Returning to (80, a) we have

$$\begin{aligned}
p'(\nu) &= x(x^2 - 3p\nu - 3l) \\
&= x\frac{\Phi(l) - 3\psi(l) - 108l^2(l^2 - a_1)^2}{36l(l^2 - a_1)^2} \\
&= \frac{x \cdot \chi}{18l(l^2 - a_1)^2}.
\end{aligned}$$

Where we define

$$\chi = \tfrac{1}{2}[\Phi(l) - 3\psi(l) - 108\,l^2(l^2 - a_1)^2]$$
$$= l^6 - 6a_1 l^4 + 4b_1 l^3 - 3a_1^2 l - b_1^2 + 4a_1^3$$
$$= A \cdot B \cdot C.\text{*})$$

Where

$$A = l^2 - (1 + k^2)l - 3k^2$$
$$B = l^2 - (1 - 2k^2)l + 3(k^2 - k^4)$$

[87] · · · · · $$C = l^2 - (k^2 - 2)l - 3(1 - k^2).$$

Refering then to note (p. 24) we have:

[88] · · · · $$p'(v) = -k^4 su^2 v \cdot cn^2 v \cdot dn^2 v = \frac{\chi(l) \cdot x}{18\,l\,(l^2 - a)^2}.$$

That is $p'(v)$ vanishes where x vanishes which gives $v = w_\lambda$ a semi-period, and in consequence, when $\Phi = 0$, f reduces to

$$f_1 = \frac{\sigma(u + w_\lambda)}{\sigma u\,\sigma(w_\lambda)}\,e^{-u\zeta(w_\lambda)} = \frac{\sigma_\alpha(u)}{\sigma(u)}. \qquad \begin{matrix}\alpha = 1, 2, 3 \\ \zeta(w_\lambda) = \eta_\lambda\end{matrix}$$

The value of the function of Lamé corresponding to any value of B giving rise to the condition $\Phi = 0$ is then deduced as follows.

From $\Phi_3 = 0$ we derive:

$$B = 5l = 1 + k^2 + 2\sqrt{19(1 - k^2)^2 + k^2}$$

and the special equation of Lamé becomes

$$y'' = [12p(u) + 1 + k^2 + 2\sqrt{19(1 - k^2)^2 + k^2}]\,y$$

and from the general form of the integral [77]

$$y = f_1'' - \tfrac{1}{5}\{1 + k^2 + 2\sqrt{19(1 - k^2)^2 + k^2}\}\,f_1.$$

But differentiating f_1 we have

$$f_1'' = [2p(u) + p(w_\lambda)]f_1$$
$$= [2p(u) + e_\alpha]f_1.$$

Hence

$$y = \left[2p(u) + e_\alpha - \tfrac{1}{5}(1 + k^2) - \tfrac{2}{5}\sqrt{19(1 - k^2)^2 + k^2}\right]\frac{\sigma_\alpha(u)}{\sigma(u)}$$

$$= \left[2pu + e_\alpha - \tfrac{1}{5}(1 + k^2) - \tfrac{2}{5}\sqrt{19(1 - k^2)^2 + k^2}\right]\sqrt{pu - e_\alpha}$$

$$= 2\left[p(u) + \tfrac{1}{2}e_\alpha - \tfrac{1}{10}B\right]\sqrt{pu - e_\alpha}$$

or

$$y = z\sqrt{pu - e_\alpha}$$

where z has the value determined by the elimentary consideration (p. 44).

---

*) Compair [161] p. 73.

Case $\chi = 0$.

If $\chi = 0$ we have a second case in which the $p'(\nu)$ vanishes, ν taking the value of a semi-period, but as this may occur without reducing x to zero the eliment will not be doubly periodic since it will contain an exponential factor e^{xu}. If then $\chi = 0$ we will have from (87) six values of B for which the integral will take the form

$$y = f_2'' - \tfrac{1}{5} B_2 f_2 \quad \text{where} \quad f_2 = \frac{\sigma(u + w_\lambda)}{\sigma u \, \sigma w_\lambda} e^{+(x - \zeta w_\lambda) \omega}$$

$$= \frac{\sigma_\lambda u}{\sigma u} e^{xu}.$$

Moreover the second integral will be

$$f_2 = \frac{\sigma_\lambda u}{\sigma u} e^{-xu}$$

the form remaining unchanged which is not as we have seen in general the case.

Case $D = 0$.

The only remaining case to be considered is where $D = 0$, or

$$l^2 - a_1 = l^2 - 1 + k^2 - k^4 = 0$$

$$\text{or} \quad l = \pm (1 - k^2 + k^4)^{1/2} = \pm \frac{\lambda}{2} \sqrt{3 g_2}$$

since $\lambda^2 g_2 = \tfrac{4}{3}(1 - k^2 + k^4)$. Also $l = 3b$ whence

$$b = \tfrac{1}{2} \sqrt{\frac{g_2}{3}}$$

$$\text{or} \quad 12 b^2 - g_2 = \varphi'(b) = 0.$$

That is $D = 0$ and $\varphi'(b) = 0$ are conditions for one and the same function of Lamé. In this case $p(\nu)$ and also the $p'(\nu)$ become infinite which gives $\nu = 0$ or the congruent values $2 m w + 2 m' w'$. The general form of our integral will not hold for this exceptional case and we are obliged to return to the treatment of the subject from the standpoint of a product.

Relation of Y and C to the Special Functions of Lamé.

Returning first to (Part IV, p. 42), the elimentary determination of the special functions of Lamé, we there found with reference to B that, first, if n be odd, it is determined by two sorts of equations, one of degree $\tfrac{1}{2}(n - 1)$ giving rise to functions of the first sort, and the other, three in all, of degree $\frac{1}{2}(n+1)$ giving rise to functions of the second sort; whence combining we have, n being odd, B determined by an equation of degree $\frac{3}{2}(n+1)+\frac{1}{2}(n-1)$ $=2n+1$. If n is even we find but one equation, degree $\frac{1}{2}n+1$, for functions of the first sort and three equations, degree $\frac{1}{2}n$, for those of the second sort making a single equation whose degree as in the first case is $2n+1$. If then these roots are all different we have in all $2n+1$ special functions of Lamé.

Returning now to the forms (65)

$$2C = \alpha'(\alpha - \beta)(\alpha - \gamma)\cdots$$

we have the half periods or values of the roots α, β that will reduce them to zero. Moreover they will not be double roots, for consider $t = e_\lambda$ as a double root of Y in which case all the terms of equation (57) will reduce to zero save the second which will be identically zero, which is a condition that the root be tripple. Differentiating we find an analogous equation and a similar course of reasoning shows that the root must be quadruple and so on which is absurde. Hence the roots that are half-periods are not double. On the other hand any other root of Y may be double but as a similar course of reasoning shows it could not be tripple.

If then $C = 0$ all the roots will be double unless they are semi-periods and we may write

[89] · · $\qquad Y = (pu - e_1)^\varepsilon (pu - e_2)^{\varepsilon'} (pu - e_3)^{\varepsilon''} \, \Pi(pu - pa)^2$

whence

[90] · · · $y = \sqrt{(pu - e_1)^\varepsilon (pu - e_2)^{\varepsilon'} (pu - e_3)^{\varepsilon''}} \, \Pi(pu - pa)$

where

$$\varepsilon,\ \varepsilon',\ \varepsilon'' = 0 \text{ or } 1.$$

But this form we observe at once is that assumed in every case by the special functions of Lamé where we found y always equal to a polynomial in $p(u)$ times some one or more of the factors $(pu - e_\lambda)^{1/2}$.

That is $C = 0$ is a condition that the integrals be the special double periodic functions of Lamé.

By a transformation similar to that on p. 35 we may write equation (64, p. 38) in the form:

$$4\,c^2 = (4\,t^3 - g_2 t - g_3) \left[\left(\frac{d\,Y}{dt}\right)^2 - 2\,Y\frac{d^2\,Y}{dt^2} \right] - (12t^2 - g_2)\,Y\frac{d\,Y}{dt}$$
$$+ 4\,[n\,(n+1)\,t + B]\,Y^2$$

and we have (62, p. 37)

$$Y = \frac{(-1)^n B^n}{[3\cdot 5\cdot 7\cdots 2n-1]^2} + \cdots$$

from which relations we see that the highest power of B in c^2 is $2n+1$ and that the condition $C = 0$ gives rise to an equation of the $2n + 1^{st}$ degree in B which is as the number of the special functions of Lamé.

Refering to (68, p. 40) we see that $C^2 = 0$ has been found as an equation of the seventh degree in B as required by the above theory.

Functions of the First Sort.

Following the notation of M. Halphen designate by P the first member of the equation that determines B corresponding to functions of the first sort. Refering again to (Part IV) we observe that if n is odd each of these functions contains the factor $p'u$.

For example we have:

$$n = 3 : y = p' \quad \text{where} \quad B = 0,$$

the degree in B being unity.

$$n = 5 : y = p''' - \tfrac{2}{3}\,Bp' = p'\,(12p - \tfrac{2}{3}\,B) \quad \text{where} \quad B^2 - 27\,g_2 = 0$$

the degree being two, etc.

But $p'(u) = 4\,(pu - e_1)\,(pu - e_2)\,(pu - e_3)$ whence for n odd or equal to three, ε, ε', ε'' are all equal to unity.

Moreover we have obtained Y (67, p. 40) expressed as a polynomial in t and b in the form

$$Y_{n=3} = \tfrac{1}{4}\,\varphi(t) - b\,[\varphi' + 3\,(t-b)^2]$$

and since

$$p'(e_\lambda) = t'(e_\lambda) = 0$$

we derive

[91] $\cdots$ $\cdots$ $\cdots$ $Y_{n=3}(e_\lambda) = -\,b\,[\varphi' + 3\,(e_\lambda - b)].$

Hence $P_{n=3} = B = 15\,b$ is a factor of $Y_{n=3}(e_\lambda)$ times a constant.

If on the other hand n be even none of the functions of the first sort contain a factor $\sqrt{pu - e_\lambda}$ and $P_{n=2\varkappa}$ will not be a factor of $Y_{n=2\varkappa}(e_\lambda)$.

Functions of the Second Sort.

We have found three equations each of degree $\frac{1}{2}(n+1)$ or $\frac{1}{2}n$ as n is taken odd or even, that give values of B that, if n be odd, correspond to functions of the second sort, or, if n be even, to functions of the third sort. Designate the first members, by Q_1, Q_2, and Q_3. Refering again to Lamé's special functions we see that if $Q_1 = 0$ the function of Lamé corresponding contains the factor $\sqrt{pu - e_1}$ if n is odd and the two corresponding factors $\sqrt{pu - e_2}$, $\sqrt{pu - e_3}$ if n is even. In the first case Q_1 is a factor of $Y(e_1)$ and in the second case of $Y(e_2)$ and of $Y(e_3)$, while in the second case we have also $Y(e_1)$ contains the factor $Q_2 Q_3$.

Returning to $n = 3$ we have (see (73) p. 44)

[92] $\quad\quad$
$$[Q_1]_{n=3} = B^2 - 6 e_1 B + 45 e_1^2 - 15 g_2$$
$$[Q_2]_{n=3} = B^2 - 6 e_2 B + 45 e_2^2 - 15 g_2$$
$$[Q_3]_{n=3} = B^2 - 6 e_3 B + 45 e_3^2 - 15 g_2$$

or in general writing $B = 15 b$ and $\varphi = \varphi(b) = 4b^3 - g_2 b - g_3$

[93] $\quad\cdots\quad$
$$[Q_\lambda]_{n=3} = 3^2 \cdot 5 \left[\varphi' + 3(e_\lambda - b)^2\right].$$

Also from (91).

[94] $\quad\cdots\quad$
$$Y(e_1) = - b \left[\varphi' + 3(e_1 - b)^2\right]$$
$$= - b \left[15 b^2 + 3 e_1^2 - 6 e_1 b - g_2\right]$$
$$= - \frac{B}{15}\left[\frac{B^2}{15} - \frac{6 e_1 B}{15} + 3 c_1^2 - g_2\right]$$
$$= - cB \left[B^2 - 6 e_1 B + 45 e_1^2 - 15 g_2\right]$$
$$= - c^2 Q_1 P$$

where in general

[95] $\quad\cdots\quad$
$$c = \frac{1}{3 \cdot 5 \cdots 2n - 1}.$$

The quantities Q are also necessarily the functions Φ times a factor as is shown by taking the substitutions

$$e_1 = \frac{1}{3\lambda}(2 - k^2), \quad e_2 = \frac{1}{3\lambda}(2k^2 - 1), \quad c_3 = -\frac{1}{3\lambda}(1 + k^2),$$
$$g^2 = \frac{4}{3\lambda}(1 - k^2 + k^4)$$

whence:

$$[Q_1]_{n=3} = B^2 - \frac{2}{\lambda}(2 - k^2)B - \frac{5 \cdot 3 k^4}{\lambda^2}$$
$$[Q_2]_{n=3} = B^2 - \frac{2}{\lambda}(1 - 2 k^2) B - \frac{5 \cdot 3}{\lambda^2}$$
$$[Q_3]_{n=3} = B^2 - \frac{2}{\lambda}(1 + k^2) B - \frac{3 \cdot 5 (1 - k^2)^2}{\lambda^2}.$$

Hence making $\lambda =$ constant, equal -1 and $B = 5l = 15b$ we obtain

$$[Q_1]_{n=3} = 5\,\Phi_1 = 5\,[5l^2 - 2\,(k^2 - 2)\,l - 3\,k^4]$$

[96] · · · $[Q_2]_{n=3} = 5\,\Phi_2 = 5\,[5l^2 - 2\,(1 - 2\,k^2)\,l - 3]$

$$[Q_3]_{n=3} = 5\cdot\Phi_3 = 5\,[5l^2 - 2\,(1 + k^2)\,l - 3\,(1 - k^2)^2].$$

Hence also:

[97] $Q = Q_1\,Q_2\,Q_3 = 5^3\,\Phi\,(l) = 5^3\,\Phi_1\,\Phi_2\,\Phi_3$

$$= 5_3\,[4\,(l^2 - a_1)^3 + (11\,l^3 - 9\,a_1\,l - b_1)^2]$$

$$= 5^3\,[125\,\xi^6 - 210\,c_1\,\xi^4 - 22\,\xi^3 + 93\,c_1{}^2\,\xi^2 + 18\,C\xi + 1 - 4\,c_1{}^3]$$

$$= \text{etc.}$$

where

$$c_1{}^3 = \frac{a_1{}^3}{b_1{}^2} = \frac{1}{108}\cdot\frac{g_2{}^3}{g_3{}^2} = 108\,\frac{(1 - k^2 + k^4)^3}{(1 + k^2)^2\,(2 - k^2)^2\,(1 - 2\,k^2)^2}.$$

We have moreover that the conditions that the integrals be special functions of Lamé are that Q_1, Q_2, Q_3 and P vanish. But $C^2 = 0$ was also found to be a condition and we note that the sum of the degrees of Q_λ and P is equal to the degree of C^2 which equals the number of the functions of Lamé. We must have then the relation

$$C^2 = c'\,Q_1\,Q_2\,Q_3\,P.$$

But we have shown that the highest power of B in the development of $4C^2$ is (p. 38)

$$4C^2 = .4BY^2 + \cdots = \frac{4B\,(-B)^{2n}}{[3\cdot 5\cdots(2n-1)]^4} + \cdots$$

whence

$$C' = \frac{1}{[3\cdot 5\cdots(2n-1)]^4} = c^4$$

which for $n = 3$ gives as before taken $c = \frac{1}{3\cdot 5}$.

We have then in general

[98] · · · · · · · · $C^2 = c^4\,PQ_1\,\dot{Q}_2\,Q_3$

and when $n = 3$

[99] · · · · · · · $C^2 = \frac{1}{(15)^4}\,QP = \frac{1}{3^4\cdot 5}\,PQ.$

If then we take

$$Q_1 = 0: B = 3e_1 \pm \sqrt{3(-12e_1^2 + 5g_2)} = k^2 - 2 \pm \sqrt{(k^2-2)^2 + 15k^4}$$

$$y = \{p + \tfrac{1}{2}e_1 - \tfrac{1}{10}\,(3e_1 \pm \sqrt{3\,(5g_2 - 12e_1^2)})\}\,\sqrt{p - e_1}$$

$$= \{p + \tfrac{1}{15}(k^2 - 2) \pm \tfrac{1}{10}\sqrt{(k^2 - 2)^2 + 15k^4}\}\sqrt{p - \tfrac{1}{3}(k^2 - 2)}$$

[100] $Q_2 = 0 : B = 3e_2 \pm \sqrt{3(5g_2 - 12e_2)} = 1 - 2k^2 \pm \sqrt{(1 - 2k^2)^2 + 15}$

$$y = \left\{ p + \tfrac{1}{2} e_2 - \tfrac{1}{10}(3e_2 \pm \sqrt{3(5g_2 - 12e_2^2)}) \right\} \sqrt{p - e_2}$$

$$= \left\{ p + \tfrac{1}{15}(1 - 2k^2) \pm \tfrac{1}{10}\sqrt{(1-2k^2)^2 + 15} \right\} \sqrt{p - \tfrac{1}{3}(1 - 2k^2)}$$

$Q_3 = 0 : B = 3e_3 \pm \sqrt{3(5g_2 - 12e_3^2)} = 1 + k^2 \pm 2\sqrt{(2 - k^2)^2 - 3k}$

$$y = \left\{ p + \tfrac{1}{2} e_3 - \tfrac{1}{10}(3e_3 \pm \sqrt{3(5g_2 - 12e_3^2)}) \right\} \sqrt{p - e_3}$$

$$= \left\{ p + \tfrac{1}{15}(1 + 2k^2) \pm \tfrac{1}{5}\sqrt{(2 - k^2)^2 - 3k} \right\} \sqrt{p - \tfrac{1}{3}(1 + k^2)}$$

all of which are special functions of Lamé of the second species, the general form being

$$y = z\sqrt{pu - e_\alpha}$$

where

$$z = p^{(n-3)} + a_1 p^{(n-5)} + \cdots + c,$$

and as given (p. 43) the general form for $n = 3$ including the above is

[101] $\qquad \qquad y = \left(p + \tfrac{1}{2} e_\alpha - \tfrac{1}{10} B \right) \sqrt{p - e_\alpha}$

where

$$B = 3e_\alpha \pm \sqrt{3(5g_2 - 12e_\alpha^2)}.$$

The Discriminant of Y.

From (65) p. 38 we have

$$2C = \alpha'(\alpha - \beta)(\alpha - \gamma) \cdots = \beta'(\beta - \alpha)(\beta - \gamma) \cdots = \cdots$$

$$= \sqrt{\varphi(a)}(\alpha - \beta)(\alpha - \gamma) \cdots \sqrt{\varphi(b)}(\beta - \alpha)(\beta - \gamma) \cdots = \cdots$$

where

$$\varphi(a) = 4(pu - e_1)(pu - e_2)(pu - e_3)$$

$$Y = (pu - e_1)^\varepsilon (pu - e_2)^{\varepsilon'}(pu - e_3)^{\varepsilon''} \Pi(pu - pa).$$

The roots of

$$\varphi(a) = 0 \quad \text{are} \quad e_1, e_2, e_3.$$

The roots of

$$Y = 0 \quad \text{are} \quad e_1, e_2, e_3, \alpha_1 \beta_1 \cdots$$

Whence the resultant of $\varphi(a)$ and Y written as the product of the differences of the roots is

$R = \Pi(\alpha - e_\lambda)$, where $\alpha = \alpha_1 \beta_1 \cdots$ to n letters and $\lambda = 1, 2$ or 3

$$= [(\alpha - e_1)(\alpha - e_2)(\alpha - e_2)][(\beta - e_1)(\beta - e_2)(\beta - e_3)] \cdots$$

$$= \frac{1}{4^n} \Pi \varphi(\alpha).$$

But again

$$Y(e_1) = [(\alpha - c_1)(\beta - e_1)(\gamma - c_1)] \cdots$$
$$Y(c_2) = [(\alpha - e_2)(\beta - e_2)(\gamma - e_2)] \cdots$$
$$Y(e_3) = [(\alpha - e_3)(\beta - e_3)(\gamma - e_3)] \cdots$$

whence

$$R = \prod_{\alpha.\lambda} (\alpha - c_\lambda) = \frac{1}{4^n} \prod \varphi(\alpha) = (-1)^n \prod_\lambda Y(e_\lambda).$$

Again we have shown (94, p. 55) that for $n = 3$; and the same method gives in general for n odd:

n odd: $Y(e_1) = -c^2 P Q_1$; $Y(e_2) = -c^2 P Q_2$; $Y(e_3) = -c^2 P Q_3$

and likewise

n even: $Y(e_1) = c^2 Q_2 Q_3$; $Y(c_2) = c^2 Q_3 Q_1$; $Y(e_3) = c^2 Q_1 Q_2$

Whence we finally derive

[102] $R = \prod_{\alpha.\lambda} (\alpha - e_\lambda) = \frac{1}{4^n} \prod \varphi(\alpha) = (-1)^n \prod_\lambda Y(e_\lambda) = \begin{cases} c^6 P^3 Q, & n\ \text{odd} \\ c^6 Q^2, & n\ \text{even.} \end{cases}$

Now the discriminant of Y equals the product of the squares of the differences of the roots and may be written:

$$\Delta = (\alpha - \beta)^2 (\alpha - \gamma)^2 \cdots$$

whence from (65)

$$\Delta^2 = \frac{2^2 C^2}{\varphi(\alpha)} \cdot \frac{2^2 C^2}{\varphi(b)} \cdots = \frac{2^{2n} C^{2n}}{\Pi \varphi(\alpha)}.$$

But we have first found

$$\Pi \varphi(\alpha) = 4^n R$$

whence

$$\Delta^2 = \frac{C^{2n}}{R}.$$

Again

$$C^2 = c^4 P Q \qquad\qquad \text{(from 99)}$$

and we derive from these n being odd

$$\Delta^2 = \frac{C^{2n}}{R} = \frac{(C^2)^n}{R} = \frac{c^{4n} P^n Q^n}{c^6 P^3 Q} = c^{2(2n-3)} P^{n-3} Q^{n-1}$$

or

$$\Delta = (-1)^{\frac{n-1}{2}} c^{2n-3} P^{\frac{n-3}{3}} Q^{\frac{n-1}{2}} : n\ \text{odd}$$

[103] and in like manner we derive   (Sign ambiguous)

$$\Delta = (-1)^{\frac{1}{2}} c^{2n-3} P^{\frac{1}{2}n} Q^{\frac{1}{2}n-1} : n\ \text{even}$$

and we have also $\Delta = 0$ since Y has at least one double root.

Case $n = 3$.

[104] $R = (\alpha - e_1)(\alpha - e_2)(\alpha - e_3)(\beta - e_1)(\beta - e_2)(\beta - e_3)(\gamma - e_1)(\gamma - e_2)(\gamma - e_3)$

$$= \frac{1}{4^3}(\alpha^3 - g_2\alpha - g_3)(\beta^3 - g_2\beta - g_3)(\gamma^3 - g_2\gamma - g_3)$$

$$= -b^3[\varphi' + 3(e_1 - b)^2][\varphi' + 3(e_2 - b)^2][\varphi' + 3(e_3 - b)^2]$$

[105] $\Delta = \frac{1}{15^3}P^0Q = \frac{(3)^3}{(15)^3}Q_1 Q_2 Q_3$

$$= 3^3[\varphi' + 3(c_1 - b)^2][\varphi' + 3(e_2 - b)^2][\varphi' + 3(e_1 - b)^2] \quad \text{(see (94))}$$

which for the special case $n = 3$ furnishes the interesting relation, Q differs only by a constant factor from the discriminant of Y.

Remembering that λ has been determined equal to (-1) we have from (97)

$$Q = 5^3[4(l^2 - a_1)^3 + (11l^3 - 9al - b)^2]$$

and the relations:

$$l = 3b : a_1 = \frac{3}{4}g_2,$$

$$b_1 = \frac{27}{4}g_3 : A_2 = \frac{1}{4}(12b^2 - g_2) : A_3 = -\frac{1}{4}(44b^3 - 3g_2 b + g_3)$$

$$4(l^2 - a_1)^3 = 4 \cdot 27 A_2^3 : 11l^3 - 9a_1 l - b_1)^2 = -27A_3$$

[106] $\cdots \quad \cdots \quad \cdots \quad Q = 3^3 \cdot 5^3[4A_2^3 + 27A_3^2]$

[107] $\cdots \quad \cdots \quad \Delta = [4A_2^3 + 27A_3^2]$

which latter value we would have derived directly from the form

$$Y = S^3 + A_2 S + A_3.$$

Writing $A_2 = \frac{1}{4}\varphi'$ and $A_3 = \frac{1}{4}\varphi - b\varphi'$ we derive still another form for Q namely

[108] $\cdot \quad \cdot \quad Q = \frac{(15)^3}{16}[\varphi'^3 + 27\varphi^2 - 8 \cdot 27 b\varphi\varphi' + 16 \cdot 27 b^2 \varphi'^2].$

Again we find

[109] $\cdots \quad x^2 = \frac{\Phi(l)}{36l(l^2 - a_1)} = \frac{4^2 3^3 \Delta}{36 \cdot 3^3 b \varphi'^2} = \left\{\frac{2}{3\varphi'}\sqrt{\frac{\Delta}{b}}\right\}^2$

$$= \left\{\frac{2}{3\varphi'}\sqrt{\frac{4A_2^3 + 27A_3^2}{b}}\right\}^2$$

from which value we again see that the vanishing of φ' is equivalent to the vanishing of D. $\quad$ (compair p. 49 and 52.)

Determination of x and v. Second Method.

We have the general theorem: every rational function of pu and $p'u$ can be written in the form:

$$\varphi_1(u) = \frac{A\,\sigma(u - v_1)\,\sigma(u - v_2)\cdots\sigma(u - v_u)}{\sigma(u - v_1')\,\sigma(u - v_2')\cdots\sigma(u - v_u')}$$

where the number of σ functions in the numerator equals the number in the denominator, making the number of zeros equal to the number of infinites. The reverse theorem is also known and we may write:

$$[110]\, (-1)^n k_1\, \frac{\sigma(u - a)\,\sigma(u - b)\,\sigma(u - c)\,\sigma(u + v)}{\sigma a\,\sigma b\,\sigma c\,(\sigma u)^4} = \Phi(pu) - \frac{1}{2C}\,p'u\,\Psi(pu)$$

where Φ and Ψ are intire polynomials in pu and $p'u$, k_1 a constant to be determined and the relation exists $a + b + c = v$. Also, from the general theory, the degree of the right hand member is four, $p(u)$ being considered as of the second degree and $p'(u)$ of the third. The degree of Φ and Ψ are thus determined as follows:

	Φ	Ψ
n odd:	$\frac{1}{2}(n + 1)$	$\frac{1}{2}(n - 3)$
n even	$\frac{1}{2}n$	$\frac{1}{2}n - 1$
$n = 3$	2	$0.$

The n roots of the first member in the general case being $a, b, c \ldots$ we have:

$$[111]\cdot \quad \cdots \quad \cdots \quad \Phi(\alpha) - \frac{1}{2C}\,\alpha'\,\Psi(\alpha) = 0$$

where

$$\alpha' = p'(\alpha), \qquad \alpha = p(\alpha).$$

From (p. 38)

$$\frac{dY}{dt} = \frac{2C}{\alpha'} = (\alpha - \beta)(\alpha - \gamma)\cdots$$

whence

$$\frac{1}{2C}\,\alpha' = \left(\frac{dt}{dY}\right)_{t=\alpha}$$

and [111] becomes

$$[112]\cdot \quad \cdots \quad \cdot \quad \left[\Phi\,\frac{dY}{dt} - \Psi\right]_{t=\alpha,\beta,\gamma,\ldots} = 0.$$

But $\alpha, \beta, \gamma, \ldots$ are also roots of Y, whence the relation

$$[113]\cdot \quad \cdots \quad \cdots \quad \Phi\,\frac{dY}{dt} - \Psi = EY$$

where E is also in general an intire polynomial in t whence

$$[114]\cdot \quad \cdots \quad \cdots \quad \cdots \quad \frac{\Phi}{Y}\,\frac{dY}{dt} = E + \frac{\Psi}{Y}$$

We have also

$$\left[\Phi\frac{dY}{dt} - \Psi\right]_{t=B} = 0$$

etc. for the other roots of Y. The degrees of [114] are

n odd: $\qquad \dfrac{\dfrac{\Psi}{Y}}{\frac{1}{2}(n-3)-n} = -\frac{1}{2}(n+3)$, $\qquad \dfrac{\Phi}{\frac{1}{2}(n+3)-1}$

n even: $\quad \frac{1}{2}(n) - 1 - n = -\left(\frac{1}{2}n+1\right)$, $\quad \left(\frac{1}{2}n+1\right) - 1$.

We have

$$Y = t^n + a_1 t^{n-1} + a_2 t^{n-2} + \cdots + a_{n-1}t + a_n$$
$$Y' = nt^{n-1} + (n-1)a_1 t^{n-2} + (n-2)a_2 t^{n-3} + \cdots + a_{n-1}$$

and

$$\frac{Y'}{Y} = \frac{nt^{n-1} + a_1(n-1)t^{n-2} + \cdots}{t^n + a_1 t^{n-1} + a_2 t^{n-2} + \cdots} = \frac{b_0}{t} + \frac{b_1}{t^2} + \frac{b_2}{t^3} + \cdots$$

or

$$nt^{n-1} + a_1(n-1)t^{n-2} + a_2(n-2)t^{n-3}$$
$$+ \cdots = b_0(t^{n-1} + a_1 t^{n-2} + \cdots) + b_1(t^{n-2} + a_1 t^{n-3} + \cdots)$$

and equating the corresponding coefficients we obtain:

[115] $\cdots \cdots \cdots \cdots$
$$\begin{aligned} b_0 &= n \\ a_1(n-1) &= na_1 + b_1 \quad \text{or} \quad b_1 = -a_1 \\ b_2 &= -2a_2 + a_1^2 \\ b_3 &= -3a_3 + a_1 a_2 - a_1^3 \\ &\text{etc.} \quad - \quad - \quad - \quad - \end{aligned}$$

Proceeding in like manner we write:

$$\Phi = B_0 t^\nu + B_1 t^{\nu-1} + \cdots + B_{n-1}t + B_\nu$$

where

$$\nu = \left[\frac{1}{2}(n+1), \frac{1}{2}n\right]$$

whence

$$\left(\frac{b_0}{t} + \frac{b_1}{t^2} + \frac{b_2}{t^3} + \cdots\right)\left(B_0 t^\nu + B_1 t^{\nu-1} + \cdots + B_\nu t + B_\nu\right)$$

$$= b_0\left(B_0 t^{\nu-1} + B_1 t^{\nu-2} + \cdots + B_{\nu-1} + \frac{B_\nu}{t}\right)$$
$$+ b_1\left(B_0 t^{\nu-2} + B_1 t^{\nu-3} + \cdots + B_{\nu-2} + B_{\nu-1}t^{-1} + B_\nu t^{-2}\right)$$
$$+ b_2\left(B_0 t^{\nu-3} + B_1 t^{\nu-4} + \cdots\right) + \cdots$$

and

$$b_0 B_\nu t^{-1} + b_0 B_{\nu-1} + b_0 B_{\nu-2} t + b_0 B_{\nu-3} t^2 + \cdots + b_0 B_1 t^{\nu-2} + b_0 B_0 t^{\nu-1}$$
$$+ b_1 B_\nu t^{-2} + b_1 B_{\nu-1} t^{-1} + b_1 B_{\nu-2} + b_1 B_{\nu-3} t + \cdots$$
$$+ - - -$$
$$+ b_{\nu-1} B_\nu t^{-\nu} + b_{\nu-1} B_{\nu-1} t^{-(\nu-1)} + \cdots + b_{\nu-1}$$
$$= b_0 B_\nu t^{\nu-1} + b_0 B_{\nu-1} t^\nu + b_0 B_{\nu-2} t^{\nu+1} + \cdots + b_0 B_1 t^{2\nu-2}$$
$$+ b_0 B_0 t^{2\nu-1} + b_1 B_\nu t^{\nu-2} + b_1 B_{\nu-1} t^{\nu-1} + b_1 B_{\nu-2} t^\nu + \cdots$$

from whence the relations:

$$b_0 B_\nu + b_1 B_{\nu-1} + b_2 B_{\nu-2} + \cdots + b_\nu B_0 = 0$$

[116]
$$b_1 B_\nu + b_2 B_{\nu-1} + b_3 B_{\nu-2} + \cdots + b_{\nu+1} B_0 = 0$$

$$\overline{\phantom{b_{\nu-1} B_\nu + b_\nu B_{\nu-1} + b_{\nu+1} B_{\nu-2}}}$$

$$b_{\nu-1} B_\nu + b_\nu B_{\nu-1} + b_{\nu+1} B_{\nu-2} + \cdots + b_{2\nu-1} B_c = 0$$

We will define:

[117]
$$\delta_m = \begin{vmatrix} b_0 b_1 & b_2 & \cdots b_m \\ b_1 b_2 & b_3 & \cdots b_{m+1} \\ - & - & - & - & - \\ - & - & - & - & - \\ b_m b_{m+1} b_{m+2} & \cdots b_{2m} \end{vmatrix}$$

We will define

$$B_0 = \delta_{\nu-1}$$

and we will then have from the above conditions, all the coefficients $B_1 B_2 \ldots$ as intire functions of $b_0 b_1 \ldots$ which are in turn functions of $a_1, a_2 \ldots$ which finally are expressed as functions of B, g_2 and g_3.

That is we have obtained Φ, of which the first coefficient shall be $\delta_{\nu-1}$ intire in terms of t, B, g_2 and g_3.

Case $n = 3$ we have:

from (p. 36)

$\mu = 2$: $2 \cdot 1 \cdot 5 \cdot 6 a_1 + 4 \cdot 3 B = 0$ or $a_1 = -\dfrac{B}{5}$

$\mu = 1$: $a_2 = \dfrac{2 B^2}{3 \cdot 5^2} - \dfrac{g_3}{4}$

$\mu = 0$: $a_3 = -\dfrac{B^3}{3^2 5^2} + \dfrac{B g_2}{3 \cdot 5} - \dfrac{g_3}{4}$

and from (115)

t^{n-1}: $n = b_0$

t^{n-2}: $a_1 (n - 1) = b_0 a_1; \quad b_1 = - a_1$

t^{n-3}: $a_2 (n - 2) = b_0 a_2 + b_1 a_1 + b_2; \quad b_2 = - 2 a_2 - b_1 a_1$

t^{n-4}: $a_3 (n - 3) = b_0 a_3 + b_1 a_2 + b_2 a_1 + b_3; \quad b_3 = 3 a_1 a_2 - 3 a_3 - a_1^3$

The conditions (116) become:

$$b_0 B_2 + b_1 B_1 + b_2 B_0 = 0$$
$$b_1 B_2 + b_2 B_1 + b_3 B_0 = 0$$

whence

$$(b_0 b_2 - b_1^2)\, B_2 = (b_1 b_3 - b_2^2)\, B_0$$
$$(b_0 b_2 - b_1^2)\, B_1 = (b_1 b_2 - b_0 b_3)\, B_0.$$

But

$$B_0 = b_0 b_2 - b_1^2 = \tfrac{3}{2}\, g_2 - \tfrac{2}{5^2}\, B^2 = \tfrac{3}{2}\, g_2 - 18 b^2 = -6 A_2 = -\tfrac{3}{2}\, \varphi'^2$$

whence

$$B_2 = (-a_1)(3 a_1 a_2 - 3 a_3 - a_1^3) - (4 a_2^2 - 4 a_1^2 a_2 + a_1^4)$$
$$= a_1^2 a_2 + 3 a_1 a_3 - 4 a_2^2$$
$$= \frac{B^4}{3^2 \cdot 5^3} + \frac{g_2 B^2}{3 \cdot 4 \cdot 5^2} + \frac{3 g_3 B}{4 \cdot 5} - \tfrac{1}{4}\, g_2^2$$
$$= 3^2 \cdot 5 b^4 + \tfrac{3}{4}\, g_2 b^2 + \tfrac{9}{4}\, g_3 b - \tfrac{1}{4}\, g_2^2$$

$$B_1 = (-a_1)(a_1^2 - 2 a_2) - 3(3 a_1 a_2 - 3 a_3 - a_1^3)$$
$$= 2 a_1^3 - 7 a_1 a_2 + 9 a_3$$
$$= -\frac{7 B^3}{3 \cdot 5^3} + \frac{g_2 B}{4} - \frac{3^2 g^3}{4}$$
$$= -3^2 7 b^3 + \tfrac{15}{4}\, g_2 b - \tfrac{9}{4}\, g_3$$
$$= 9 A_3 + 12 b A_2 = \tfrac{9}{4}\, \varphi - 6 b \varphi'.$$

We derive then finally

$$\Phi = B t^2 + B_1 t + B_2$$
$$= -6\left(3 b^2 - \tfrac{1}{4}\, g_2\right)(S_2 + 2 b S + b^2)$$
$$\qquad + \left(63 b^3 + \tfrac{15}{4}\, g_2 b - \tfrac{9}{4}\, g_3\right)(S + b)$$
$$\qquad + 3^2 \cdot 5^2 b^4 + \tfrac{3}{4}\, g_2 b^2 + \tfrac{9}{4}\, b g_3 - \tfrac{1}{4}\, g_2^2.$$

Coef. S^2 is $-6\left(3 b^2 - \tfrac{1}{4}\, g_2\right) = -6 A_2$

„ S is $-9\left(-11 b^3 + \tfrac{3}{4}\, g_2 b - \tfrac{1}{4}\, g_3\right) = 9 A_3$

„ S^0 „ $-4\left(3 b^2 - \tfrac{1}{4}\, g_2\right)^2 = -4 A_2^2$

Hence

[118] $\Phi = -6\left(3 b^2 - \tfrac{1}{4}\, g_2\right) S^2 + 9\left(-11 b^3 + \tfrac{3}{4}\, g_2 b - \tfrac{1}{4} g_3\right) S - 4\left(3 b^2 - \tfrac{1}{4}\, g_2\right)^2$

$$= -6 A_2 S^2 + 9 A_3 S - 4 A_2^2$$

Having obtained Φ, the calculation of Ψ and E is simplified by the following considerations:

Let n be taken odd and take for B a root of the equation $Q_1 = 0$. In this case (see p. 54) we have Y as a product of $t - e_1$ by a polynomial U^2 where U has the degree $\frac{1}{2}(n-1)$. Moreover U enters as a double factor and is therefore also a factor of Y', whence, from the form

$$\Phi \frac{dY}{dt} - \Psi = EY$$

we find that U must also be a factor of Ψ. This, however, we know to be impossible since the degree of U is $\frac{1}{2}(n-1)$ and that of Ψ only $\frac{1}{2}(n-3)$ (p. 60).

The only conclusion possible then is that Ψ contains a zero factor. We know also that B being any value whatever, Ψ considered as a function of B contains the factors Q_1, Q_2 and Q_3 and it follows that we may write

$$\Psi_{n \text{ odd}} = Q\Theta$$

where $Q = 0$, if B be taken as a root of $Q_1 = 0$, $Q_2 = 0$, or $Q_3 = 0$, and $\Theta = f_\Theta(t)$.

By a similar course of reasoning we show that if B be taken as a root of P, n being even, Y will be the square of a polynomial V of degree $\frac{1}{2}n$ where Ψ is only of degree $\frac{1}{2}n - 1$, and that in consequence one has

$$\Psi_{n \text{ even}} = P\Theta$$

Hence we write:

[119] · · · · · · · $\begin{cases} n \text{ odd:} \quad \Phi \dfrac{dY}{dt} - Q\Theta = EY \\[2mm] n \text{ even:} \quad \Phi \dfrac{dY}{dt} - P\Theta = EY \end{cases}$

where all the functions are intire in t.

As we have before determined the first coefficient of Φ is the determinant $\delta_{\nu-1}$ and in like manner we find the first coefficient of Ψ to be

[120] · · · · · $\delta_\nu = b_\nu B_\nu + b_{\nu+1} B_{\nu-1} + \cdots + b_{2\nu} B_0.$

Hence if we divide δ_ν by Q, n being odd we will have γ, the first coefficient of Θ.

To find E, $n = 3$.

The degree of Φ is $\frac{1}{2}(n+1)$, the degree of Y is $n-1$ and the degree of Ψ is $\frac{1}{2}(n-3)$ less than Y'. Hence from the relation on $p\,(64)$, the degree of EY must be

$$\tfrac{1}{2}(n+1) + (n-1) = \tfrac{1}{2} \cdot (3n-1).$$

But the degree of Y is n and hence the degree of E is $\frac{1}{2}(n-1)$. We have then

[121]
$$E_{n=3} = \eta t + \eta_1$$

and Ψ reduces to a constant, namely:

[122]
$$\Psi_{n=3} = \gamma Q$$

We have:
$$Y_3 = S^3 + A_2 S + A_3$$
$$Y_3' = 3 S^2 + A_2$$
$$\Phi = -6 A_2 S^2 + 9 A_3 S - A_2{}^2$$

and substituting we derive

$$(3 S^2 + A_2)(-6 A_2 S^2 + 9 A_3 S - 4 A_2{}^2) = (\eta S + \eta_1)(S^3 + A_2 S + A_3) + \gamma Q$$

and from these we have

$$\eta = -18 A_2 ; \qquad \eta_1 = 27 A_3$$

whence

[123]
$$E_3 = -9\,[2 A_2 S - 3 A_3].$$

Returning to our original form we find that when n is three we may write:

[124]
$$(-1)^n k_1 \frac{\sigma(u-a)\,\sigma(u-b)\,\sigma(u-c)\,\sigma(u+v)}{\sigma a\,\sigma b\,\sigma c\,(\sigma u)^4} = \Phi(pu) - \frac{1}{2c}\,p'u\,\Psi(pu)$$

$$= \Phi - \frac{1}{2c} S' \gamma Q = (-6 A_2 S^2 + 9 A_3 S - 4 A_2{}^2) + \frac{1}{2C} S'(4 A_2{}^3 + 27 A_3{}^2).$$

Having this development, the determination of x and v is made possible as follows: — Taking the derivative of the log. of the first member, A, and developing according to the powers of u we write in general

$$A = c_1\,[\zeta(u-a) + \zeta(u-b) + \zeta(u-c) \cdots \zeta(u+v) - (n+1)\zeta u.$$

But the developments are known:

$$\zeta(u+v) - \zeta u = \zeta(v) - \frac{1}{u} - up(v) - \frac{u^2}{2}\,p'v \cdots$$

$$\zeta(u-a) - \zeta(u) = -\zeta a - \frac{1}{u} - upa - \frac{u^2}{2}\,p'a \cdots$$

$$\zeta(u-b) - \zeta(u) = -\zeta b - \frac{1}{u} - upb - \frac{u^2}{2}\,p'b \cdots$$

5

and we may write

$$A = (\zeta v - \zeta a - \zeta b - \zeta c \cdots) - \frac{n+1}{u} - (pv + \alpha + \beta + \gamma + \cdots)u$$
$$- \frac{u^2}{2}(p'v + p'a + p'b + \cdots)\cdots$$

But

$$\zeta v - \zeta a - \zeta b - \zeta c - = x \quad \text{and} \quad p'a + p'b + p'c + \cdots = 0$$

(see pages 25 and 45), whence

[125] $$A = -\frac{n+1}{u} + x - (\alpha + \beta + \gamma + \cdot + pv)\cdot u - \frac{u^2}{2}p'v + \cdots$$

The degree of Φ is $\frac{1}{2}(n+1)$, of S', 3, of Ψ, $\frac{1}{2}(n-3)$, and of p', $\frac{3}{2}$ which gives the degree of the second member as $\frac{1}{2}(n+1)$, also

$$pu = \frac{1}{u^2} + \cdots \quad \text{whence} \quad p^{\frac{1}{2}(n+1)} = \frac{1}{u^{2\left(\frac{1}{2}n+1\right)}} + \cdots = \frac{1}{u^{n+1}} + \cdots$$

and developing the second member (B) we write, disregarding the constant factor

$$B = \frac{1}{u^{n+1}} + \frac{q_1}{u^n} + \frac{q_2}{u^{n-1}} + \frac{q_3}{u^{n-2}} + \cdot$$

whence

$$[126]\, d \log B = -\frac{\dfrac{n+1}{u^{n+2}} - \dfrac{nq_1}{u^{n+1}} - \dfrac{q_2(n-1)}{u^n} - \dfrac{(n-2)q_3}{u^{n-1}} -}{\dfrac{1}{u^{n+1}} + \dfrac{q_1}{u^n} + \dfrac{q_2}{u^{n-1}} + \dfrac{q_3}{u^{n-2}} + \cdots}$$

$$= -\frac{1}{u} \cdot \frac{(n+1) + nuq_1 + (n+1)q_2 u^2 + (n-2)q_3 u^3 + \cdots}{1 + q_1 u + q_2 u^2 + q_3 u^3 + \cdot}$$

$$= -\frac{n+1}{u} + q_1 + (2q_2 - q_1^2)u + (3q_3 - 3q_1 q_2 + q_1^3)u^2 + \cdots$$

Again:

$$\Phi = B_0 t^{\frac{1}{2}(n+1)} + B_1 t^{\frac{1}{2}(n-1)} + \cdots$$

$$\Theta = \gamma t^{\frac{1}{2}(n-3)} + \gamma_1 t^{\frac{1}{2}(n-5)} + \cdots$$

$$p' = (4t^3 - tg_2 + g_3)^{\frac{1}{2}} = -\frac{2}{u^3} + \cdots$$

whence

$$B = \left(B_0 t^{\frac{1}{2}(n+1)} + B_1 t^{\frac{1}{2}(n-1)} + \cdots\right) - \frac{Q}{2C}(4t^3 - tg_2 - g_3)^{\frac{1}{2}}\left(\gamma t^{\frac{1}{2}(n-3)} + \gamma t^{\frac{1}{2}(n-5)} + \cdots\right)$$

$$= \frac{B_0}{u^{n+1}} + \frac{Q\gamma}{Cu^n} + \frac{B_1}{u^{(n-1)}} + \frac{Q\gamma_1}{Cu^{n-2}} + \cdots$$

and

$$[127]\quad d\log. B = B_0\left[-\frac{n+1}{u} + \frac{Q\gamma}{C} + \left(2B_1 - \frac{Q^2\gamma^2}{C^2}\right)u + \left(\frac{3Q\gamma_1}{C} - \frac{3Q\gamma}{C}B_1 + B_1^3\right)\right]u^2$$

From developments [126] and [127] we find

$$[128]\quad \cdots\quad q_1 = \frac{Q\gamma}{CB_0};\quad q_2 = \frac{B_1}{B_0};\quad q_3 = \frac{Q\gamma_1}{CB_0}, \quad n\ being\ odd,$$

and from developments [125] and [126]

$$[129]\quad \begin{cases} x = q_1 = \dfrac{Q\gamma}{CB_0} \\[2mm] \alpha + \beta + \gamma + \cdots + p\,\nu) = q_1^2 - 2q_2 = \dfrac{Q^2\gamma^2}{C^2B_0^2} - \dfrac{2B_1}{B_0} \\[2mm] p'\nu = 2\left(3q_1q_2 - 3q_3 + q_1^3\right) = \dfrac{6Q\gamma B_1}{CB_0^2} - \dfrac{6Q\gamma_1}{CB_0} + \dfrac{2Q^3\gamma^3}{C^3B_0^3}. \end{cases}$$

These forms are transformed by the aid of the relations

$$C = c^2\sqrt{PQ}\ (p.\ 56);\quad (2n-1)(\alpha + \beta + \gamma + \cdots) = B\ (p.\ 29)$$

$$(2n-1)a_1 = -B\ (p.\ 36)\ \text{whence}\ (\alpha + \beta + \gamma + \cdots) = -a_1 = \frac{B}{2n-1}$$

giving as result:

$$[130]\quad \cdots\quad \begin{cases} x = \dfrac{Q\gamma}{CB_0} = \dfrac{\gamma}{c^2B_0}\sqrt{\dfrac{Q}{P}} \qquad\qquad n\ odd. \\[2mm] p\nu = \dfrac{Q\gamma^2}{c^4PB_0^2} - \dfrac{2B_1}{B_0} - \dfrac{B}{2n-1} \\[2mm] p'\nu = -\dfrac{2}{c^2}\left\{\dfrac{Q\gamma^3}{c^4PB_0^3} - \dfrac{3\gamma B_1}{B_0^2} + \dfrac{3\gamma_1}{B_0}\right\}\sqrt{\dfrac{Q}{P}} \\[2mm] \text{and from these the combined forms arise} \\[2mm] \dfrac{p'\nu}{2x} = -\dfrac{Q\gamma^2}{c^4PB_0^2} + \dfrac{3B_1}{B_0} - \dfrac{3\gamma_1}{\gamma} \\[2mm] \dfrac{p'\nu}{2x} + p\nu = \dfrac{B_1}{B_0} - \dfrac{3\gamma_1}{\gamma} - \dfrac{B}{2n-1} \end{cases}$$

These formules are perfectly general for n odd and the corresponding forms n even obtained in like manner are

$$[131]\quad \cdots\quad \begin{cases} x = \dfrac{CB_0}{P\gamma} = \dfrac{c^2B_0}{\gamma}\sqrt{\dfrac{Q}{P}} \qquad\qquad n\ even. \\[2mm] p\nu = \dfrac{c^4QB_0^2}{P\gamma^2} = \dfrac{2\gamma_1}{\gamma} - \dfrac{B}{2n-1} \\[2mm] p'\nu = -2c^2\left\{\dfrac{c^4QB_0^3}{P\gamma^3} - \dfrac{3B_0\gamma_1}{\gamma^2} + \dfrac{3B_1}{\gamma}\right\}\sqrt{\dfrac{Q}{P}} \\[2mm] \dfrac{p'\nu}{2x} + p\nu = \dfrac{\gamma_1}{\gamma} - \dfrac{2B_1}{B_0} - \dfrac{B}{2n-1}. \end{cases}$$

The superiority of these forms over those first derived, showing as they do at a glance the synthetic relations, is unquestionable and the explicit forms for our case n equals three and also for n equals four and to some extent for yet higher values, are obtainable with greater easy than by the first method. Even here however the forms increase in complexity so rapidly that n is practically restricted to the lowest values.

For case $n = 3$.

We have found all the eliments except γ_1 which is derived from development of Θ, or more easily as follows.

From (106, p. 59)

$$Q = (15)^3 \left[4 A_2^3 + 27 A_3^2 \right]$$

and from (p. 65)

$$\Psi_{n=3} = Q\gamma = (3 S^2 + A)\left(- 6 A_2 S^2 + 9 A_3 S - 4 A_2^2 \right)$$
$$+ 9 (2 A_2 S - 3 A_3)(S^3 + A_2 S + A_3)$$
$$= - \left(4 A_2^3 + 27 A_3^2 \right)$$

and a comparison gives immediately

[132] ·   ·   ·   · $$\gamma = - \frac{1}{(15)^3}.$$

The other values for the eliments have been found, namely:

$$c = \tfrac{1}{15} \qquad\qquad \gamma_1 = 0 \qquad\qquad \varphi' = 12 b^2 - g_2$$

$$P = 15 b \qquad\qquad A_2 = \tfrac{1}{4}\,\varphi' \qquad\qquad l = 3 b$$

$$B_0 = - \tfrac{3}{2}\,\varphi' \qquad A_3 = \tfrac{1}{4}\,\varphi - b\varphi' \qquad a_1 = \tfrac{3}{4}\,g_2$$

$$B_1 = \tfrac{9}{4}\,\varphi - 6 b\varphi' \qquad \varphi = 4 b^3 - b g_2 - g_3 \qquad b_1 = - \tfrac{27}{4}\,g_3.$$

We have then for n equals three

[133] · $$x = \frac{\gamma}{c^2 B_0} \sqrt{\frac{Q}{P}} + \frac{2}{(15)^3} \cdot \frac{(15)^2}{3\,\varphi'} \sqrt{\frac{(15)^3 \left(4 A_2^3 + 27 A_3^2 \right)}{15 b}}$$

$$= \frac{2}{3\,\varphi'} \sqrt{\frac{4 A_2^3 + 27 A_3^2}{b}}$$

(compair 109, p. 59.)   ·

$$= \frac{1}{6\,\varphi'} \left\{ \frac{\varphi'^3 + 27\,\varphi^2 - 8 (27)\, b\,\varphi\,\varphi' + 16 (27) b^2 \varphi'^2}{b} \right\}^{\tfrac{1}{2}}$$

Squaring we have:

[134].

$$x^2 = \frac{4\left(3b^2 - \frac{1}{4}g_2\right)^3 + 27\left(11b^3 - \frac{3}{4}bg_2 + \frac{1}{4}g_3\right)^2}{36b\left(3b^2 - \frac{1}{4}g_2\right)^2}$$

$$= \frac{4(l^2 - a_1)^3 + \left(11b_1^3 - 9a_1b_1 - b_1\right)^2}{36l(l^2 - a_1)^2}$$

$$= \frac{\Phi(l)}{36l(l^2 - a_1)^2}$$ (compair 82, p. 49.)

$$= \frac{\Phi_1\Phi_2\Phi_3}{36l(l^2 - a_1)^2} = \frac{Q_1Q_2Q_3}{5^36^2l(l^2 - a_1)^2} = \frac{PQR}{SD^2} = \text{etc.}^*)$$

Again we have:

[135].

$$pv = \frac{Qy^2}{c^4PB_0^2} - \frac{2B_1}{B_0} - \frac{B}{2n-1}$$

$$= \frac{4\left[4A_2^3 + 27A_3^2\right]}{9\varphi'^2b} + \frac{4\left(\frac{9}{4}\varphi - 6b\varphi'\right)}{3\varphi'} - 3b$$

whence

[136] $pv - b = \dfrac{4\left[\frac{1}{16}\varphi'^3 + 27\left(\frac{1}{16}\varphi^2 - \frac{1}{2}b\varphi\varphi' + b^2\varphi'^2\right)\right] + \frac{108}{4}\varphi\varphi'b - .72b^2\varphi'^2}{9\varphi'^2b}$

$$= \frac{\varphi'^3 + 27\varphi^2 - 108b\varphi\varphi'}{36b\varphi'^2}.$$

Writing φ and φ' in terms of $g_2 \cdot g_3$ and b we have:

$$\varphi'^3 = 1728b^6 - 432b^4g_2 + 36b^2g_2^2 - g_2^3$$

$$27\varphi^2 = 432b^6 - 216b^4g_2 + 27b^2g_2^2 - 216b^3g_3 + 27g_3^2 + 54g_2g_3b.$$

$$-108b\varphi\varphi' = -5184b^6 + 1728b^4g_2 - 108b^2g_2^2 + 1296b^3g_3 - 108bg_2g_3$$

whence

[137] $pv = \dfrac{2160b^6 + 216b^4g_2 + 1080b^3g_3 - 9b^2g_2^2 - 54bg_2g_3 - g_2^3 + 27g_3^2}{36b\left(144b^4 - 24b^2g_2 + g_2^2\right)^2}$

Again from the first method

$$pv = -\Omega = \frac{1+k^2}{3} - k^2sn^2v = \frac{\psi(3b)}{108b(9b^2 - a_1)^2}$$

where

$$\psi = 5(3b)^6 + 6a_1(3b)^4 - 10b_1(3b)^3 - 3a_1^2(3b)^2 + 6a_1b_1(3b) + b_1^2 - 4a_1^3$$

or expanding we again obtain

[138] $pv = \dfrac{2160b^6 + 216b^4g_2 + 1080g_3b^3 - 9b^2g_2^2 - 54bg_2g_3 - g_2^3 + 27g_3^2}{36b\left(144b^4 - 24b^2g_2 + g_2^2\right)^2}$

$$= \frac{\varphi'^3 + 27\varphi^2 - 108b\varphi\varphi' + 36b^2\varphi'^2}{36b\varphi'^2}.$$

---

*) Compair Hermite where $P = \Phi_1$, $Q = Q_2$, $R = Q_3$, $S = 36l$, $D = (l^2 - a)$, $a = a_1$.

It is, finally, evident from the general forms that if it be required to determine $p'v$ it will be easier first to find

$$\frac{p'v}{2x} + pv = \frac{B_1}{B_0} - \frac{3\gamma_1}{\gamma} - \frac{B}{2n-1}$$

$$= \frac{\frac{9}{4}\varphi - 6b\varphi'}{\frac{3}{2}\varphi'} - 3b = \frac{2b\varphi' - 3\varphi}{2\varphi'}$$

$$= b - \frac{3}{2}\frac{\varphi}{\varphi'}$$

Whence

$$p'v = \left(b - \frac{3}{2}\frac{\varphi}{\varphi'} - pv\right)2x = -\left\{(pv - b) + \frac{3}{2}\frac{\varphi}{\varphi'}\right\}2x$$

$$= \frac{162\,b\varphi\varphi' - 27\varphi^2 - \varphi'^3}{108\,\varphi'^3 b^{\frac{1}{2}}}\sqrt{\varphi'^3 + 27\varphi^2 - 216\,b\varphi\varphi' + 432\,b^2\varphi'^2}.$$

Determination of v.   Third Method.

The formulae may be obtained by a third method and in yet different forms as follows:

Starting anew with equation [110] we write

[139] $(-1)^n k_1 \dfrac{\sigma(u-a)\,\sigma(u-b)\cdots\sigma(u+v)}{\sigma a\,\sigma b\cdots\sigma v\,(\sigma u)^{n+1}} = \Phi(pu) - \dfrac{1}{2c}p'u\,\Psi(pu).$

Also

$$\left[\frac{\sigma(u-a)}{\sigma a}\right]_{u=0} = -1$$

whence it follows that the left hand member of (139) depends for its value on the terms

$$\frac{(-1)^n k_1}{\sigma(u)^{n+1}}.$$

But we have again

$$\left[\frac{\sigma u}{u}\right]_{u=0} = 1$$

whence we may write, taking n odd

$$\left[(-1)^n k_1 \frac{\sigma(u+a)\cdots\sigma(u+v)}{\sigma(a)\,\sigma(b)\cdots\sigma(v)\,(\sigma u)^{n+1}}\right]_{u=0} = \frac{k_1}{u^{n+1}} + \cdots$$

and from p. 66

$$= \frac{B_0}{u^{n+1}} + \frac{\varphi\gamma}{C}\cdot\frac{1}{u^n} + \cdots$$

That is n being odd

$$k_1 = B_0.$$

And a similar investigation gives n being even

$$k_1 = \frac{P\gamma}{C}.$$

Since $\nu = a + b + c$ we may write

$$e^{-a+b+c+\cdots-\nu)}\eta_1 = 1$$

and multiplying by this factor we can separate the left hand member into factors of the form

[140]. . . . . $\dfrac{\sigma(a+u)}{\sigma u\,\sigma a}\,e^{-a}\eta_1 = \dfrac{\sigma_1 a}{\sigma a}$

for $u = w_1$.

But for this value $p'(w_1) = 0$ and our relation becomes

$$k\,\frac{\sigma_1 a\,\sigma_1 b\cdots\sigma_1 \nu}{\sigma a\sigma b\cdots\sigma\nu} = \Phi(pw_1) = \Phi(e_1) = \Phi_1.$$

[141] And we obtain in a similar manner

$$k\,\frac{\sigma_2 a\,\sigma_2 b\cdots\sigma_2 \nu}{\sigma a\sigma b\cdots\sigma\nu} = \Phi(pw_2) = \Phi(e_2) = \Phi_2$$

and

$$k\,\frac{\sigma_3 a\,\sigma_3 b\cdots\sigma_3 \nu}{\sigma a\sigma b\cdots\sigma\nu} = \Phi(pw_3) = \Phi(e_3) = \Phi_3.$$

Recalling the known relation

$$p'u = -2\,\frac{\sigma_1 u\,\sigma_2 u\,\sigma_3 u}{\sigma^3 u}$$

we have upon taking the product of the above equations

[142]. .   . $k^3 p'a\,p'b\cdots p'\nu = (-2)^{n+1}\,\Phi_1\,\Phi_2\,\Phi_3.$

Again from the relations (65)

$$\alpha' = \frac{2C}{(\alpha-\beta)(\alpha-\gamma)(\alpha-\delta)\cdots}\quad\text{etc.}$$

to n terms and we obtain the product

[143] $\alpha'\beta'\gamma'\cdots = \dfrac{2^n c^n(-1)^{1\cdot2\cdots n-1)}}{(\alpha-\beta)^2(\alpha-\gamma)^2(\alpha-d)^2\cdots(\beta-\gamma)^2(\beta-d)^2\cdots(\gamma-d)^2\cdot} = \dfrac{(-1)^{\frac{1}{2}n(n-1)}(2)^n C^n}{}$

$$= \frac{(-1)^{\frac{1}{2}n(n-1)}(2)^n C^n}{\Delta}$$

Δ being the discriminant of Y. Substituting this value in [142] we derive

[144]. . . $(-1)^{\frac{1}{2}n(n-1)}2^n C^n p'\nu = (-1)^{n+1}2\,\Phi_1\Phi_2\Phi_3\Delta.$

Again squaring we get

$$k^2\,\frac{\sigma_1^2 a\,\sigma_1^2 b\cdots\sigma_1^2\nu}{\sigma^2 a\,\sigma^2 b\cdots\sigma^2\nu} = \Phi^2(e_1) = (-1)^n k^2(pa-e_1)(pb-e_1)\cdots(p\nu-e_1)$$

or (see [89])

[145]. . . . . $(-1)^n k^2\,Y(e_1)\,(p\nu - e_1) = \Phi^2(e_1).$

and we have also the two corresponding expressions.

We have shown (see p. 58) that when $t = e_1$ we have $Y(e_1) = -c^2 PQ$ whence it follows from this and relation [145] that $\Phi(e_1)$ is divisable by Q_1 and in general $\Phi(e_\lambda)$ by Q_λ. We thus derive the relations

[146] $\cdots \quad \Phi_1 = Q_1 F_1 \;:\; \Phi_2 = Q_2 F_2 \;:\; \Phi_3 = Q_3 F_3.$

We have also found n being odd:

$$k = B_0 : C = c^2 \sqrt{PQ} : \varDelta = (-1)^{\frac{n-1}{2}} c^{2n-3} P^{\frac{1}{2}(n-3)} Q^{\frac{1}{2}(n-1)}$$

$$Y(e_1) = -c^2 P Q_1$$

These values in [144] give

$$(-1)^{\frac{1}{2}(n-1)n} B_0^3 c^{2n} P^{\frac{n}{2}} Q^{\frac{n}{2}} p'\nu$$

$$= (-1)^{(n+1)+\frac{n-1}{2}} 2 c^{2n-3} P^{\frac{1}{2}(n-3)} Q^{\frac{1}{2}(n-1)} Q F_1 F_2 F_3$$

or

[147] $\cdots \quad p'\nu = \dfrac{2 Q F_1 F_2 F_3}{c^3 P^{\frac{3}{2}} B_0^3 Q^{\frac{1}{2}}} = \dfrac{2 F_1 F_2 F_3}{c^3 P B_0^3} \sqrt{\dfrac{Q}{P}}$

and from [145]

$$B_0^2 c^2 P Q_1 (p\nu - e_1) = \Phi_1^2 = Q_1^2 F_1^2 \qquad\qquad n \text{ odd.}$$

Whence we have in general

[148] $\cdots\cdots\cdots \quad p\nu - e_\lambda = \dfrac{Q_\lambda F_\lambda^2}{c^2 B_0^2 P}.$

The corresponding expressions for n even are

[149] $\cdots\cdots\cdots \begin{cases} p'\nu = -\dfrac{2 c^3 \Phi_1 \Phi_2 \Phi_3}{\gamma^3 P} \cdot \sqrt{\dfrac{Q}{P}} \\[2mm] p\nu - e_\lambda = \dfrac{c^2 Q_\lambda \Phi_\lambda^2}{\gamma^2 P} \end{cases}$

Again from [130]

$$p'\nu = -\frac{2}{c^2} \left\{ \frac{Q\gamma^3}{c^4 P B_0^3} - \frac{3 \gamma B_1}{B_0^2} \right\} \sqrt{\frac{Q}{P}}$$

$$= \frac{2 F_1 F_2 F_3}{c^3 B_0^3 P} \sqrt{\frac{Q}{P}}$$

$$= -\frac{2}{c^3 B_0^3 P} \left\{ \frac{Q\gamma^3}{c^3} - \frac{3 \gamma B_1 c^4 B_0 P}{c^3 B_0^3 P} \right\} \sqrt{\frac{Q}{P}}$$

$$= -\frac{2\gamma}{c^3 B_0^3 P} \left\{ \frac{Q\gamma^2 - 3 B_0 B_1 P c^4}{c^3} \right\} \sqrt{\frac{Q}{P}}.$$

Compairing the second and fourth forms we have

[150] $\cdots\cdots \quad F_1 F_2 F_3 = -\dfrac{1}{c^3}(Q\gamma^2 - 3 B_0 B_1 P c^4).$

Substituting the values $n = 3$ (p. 68) and refering to the value of χ (p. 51) we find the relation

[160]· · · · $F_{n=3} = [F_1 F_2 F_3]_{n=3} = \dfrac{8}{3^3 15^3}\,\chi = \dfrac{8}{3^0 5^3}\,ABC.$

It follows then that χ, if expressed in terms of the modulus k and b or as a function of b, e_λ, g_2 and g_3, will be separable into three factors which from the expressions for Φ are seen to be of the same degree in b, namely, the second.

The factors of χ which we before obtained by inspection (see p. 51 [87]) are

[161]· · · $\begin{aligned} A &= l^2 - (1 + k^2)\,l - 3k^2 \\ B &= l^2 - (1 - 2k^2)\,l + 3(k^4 - k^4) \\ C &= l^2 - (k^2 - 2)\,l - 3(1 - k^2) \end{aligned}$

and we find the relations:

[162]· · · · · $F_1 = \tfrac{2}{45}A; \quad F_2 = \tfrac{2}{45}B; \quad F_3 = \tfrac{2}{45}C.$

Taking now $S = 36l$ and $D = l^2 - a_1 = l^2 - 1 + k^2 - k^4$ we find the following relations of M. Hermite

[163]
$$\left\{\begin{aligned}
& x^2 = \frac{\Phi(l)}{36\,l\,(l^2 - a_1)} = \frac{PQR}{SD^2} \\[1mm]
& -p'v = \Omega_1 = k^2\,snu\,cnu\,dnu = \frac{-\chi(l)x}{36\,l\,(l^2 - a_1)} = -\frac{ABCx}{SD^2} \\[1mm]
& k^2 sn^2\omega = \frac{1 + k^2}{3} - \frac{\psi}{36\,l(l^2 - a_1)^2} = \frac{12\,l(l^2 - a_1)^2(1 + k^2) - \psi(l)}{36\,l(l^2 - a_1)^2} = -\frac{PA^2}{SD^2} \\[1mm]
& \text{whence} \\[1mm]
& k^2 cn^2\omega = \frac{12\,l(l^2 - a_1)^2(2k^2 - 1) + \psi(l)}{36\,l(l^2 - a_1)^2} = \frac{QB^2}{SD^2} \\[1mm]
& dn^2\omega = \frac{12\,l(l^2 - a_1)^2(2 - k^2) + \psi(l)}{36\,l(l^2 - a_1)^2} = \frac{RC^2}{SD^2}
\end{aligned}\right.$$

where $x = \lambda$ and $a_1 = a$ and $\omega = v$. (see also note p. 69)

General Discussion.

Reviewing the foregoing theory we have found that when $n = 3$

$$y = f'' - 3bf$$

and that in general y is a function of f where we write

$$f = \frac{\sigma(u + v)}{\sigma u}\,e^{(x - \zeta v)u}$$

the one exception occurring where v equals zero.

We find further, that where Q or Φ vanish in which case x and $p'\nu$ also vanish, our integrals, six in number ($n = 3$), become doubly periodic and are in fact the original special functions of Lamé of the second and third sort.

We have found for x the general value

$$x = \frac{\gamma}{c^z B_0} \sqrt{\frac{Q}{P}}$$

from which form we see that x will be zero when γ and Q vanish and will be infinite where B or P vanish. But from the form

$$p\nu = \frac{Q\gamma^2}{c^4 P B_0{}^2} - \frac{2B_1}{B} - \frac{B}{2n+1}$$

we observe that $p\nu$ is also infinite where x becomes infinite through the vanishing of B_0.

We have further that in case P vanish the integral becomes a function of Lamé of the first sort in which p takes the place of f in the general solution the form being

$$[164]\ (-1)^n y = \frac{1}{(n-1)!} p^{(n-2)} u + \frac{1}{(n-3)!} q_2 p^{(n-4)} u + \frac{1}{(n-5)!} q_4 p^{(n-6)} u + \cdots$$

the values of B conforming with the above cases being roots of the equations $P = 0$, $Q_1 = 0$, $Q_2 = 0$, $Q_3 = 0$.

Moreover when Q vanishes x and $p'\nu$ will vanish simultaniously which makes ν one of the semi-periods $\omega\lambda$, and f may be written

$$[165] \cdot \quad \cdot \quad \cdot \quad \cdot \quad \cdot \quad f_{Q=0} = \pm \frac{\sigma_\lambda u}{\sigma u}.$$

Again, observing the last forms obtained, we see that ν can also be a half period if F_λ, n being odd, or Φ_λ, n being even, vanish, but it does not follow that x will also reduce to zero. That is the integral will in general have the form

$$[166] \cdot \quad \cdot \quad \cdot \quad \cdot \quad f_1 = \frac{\sigma(u + w_\lambda)}{\sigma u} e^{(x - \zeta(w_\lambda))u} = \frac{\sigma_\lambda u}{\sigma u} e^{xu}$$

when

$$F_\lambda = 0, \text{ or } \Phi_\lambda = 0, \text{ or } \chi = 0, \text{ or } A = 0, \text{ or } B_0 = 0, \text{ or } C = 0.$$

In this case as in general two distinct integrals exist which are doubly periodic of the second species the second integral being

$$f_2 = \frac{\sigma_\lambda u}{\sigma u} \rho^{-xu}$$

a form which does not differ from f, a peculiarity which does not appear in the special functions of Lamé.

We have finally but one more case to consider, namely when $\nu = 0$, a condition arising when B_0 or γ, common to the functions x, $p\nu$ and $p'\nu$, vanish, in which case the integrals become functions named after their discoverer.*)

Functions of M. Mittag-Leffler.

As M. Hermite observes (p. 28) the vanishing of A, B, C and D are necessary conditions that the integrals shall be functions which he first called functions of M. Mittag-Leffler, but they are not sufficient conditions. The functions are in fact special cases of f_1 and f_2 having the additional property that the logarithms of the so called multiplicators are proportional to the corresponding periods. In this case the integrals assume a special form where the elimentary function is a function of p and p' multiplied by a determinate exponential having the above property. We can show that these are but special cases of the general doubly periodic function of the second species of M. Hermite as follows:

We have as the general form

[167] $\quad F(u) = \dfrac{\sigma(u - a_1)\,\sigma(u - a_2) \ldots \sigma(u - a_{n-1})}{\sigma(u - b_1)\,\sigma(u - b_2) \ldots \sigma(u - b_{n-1})}\, e^{\varrho u}$ (see (16) p. 17.)

and as a function of the second species upon the addition to the arguments of the periods $2w$ and $2w'$ the function remains unchanged save in the exponential factor which takes the forms respectively

[170] . . . . . $\quad \begin{aligned} \mu &= e^{2\eta}(B - A) + 2\varrho w \\ \mu' &= e^{2\eta'}(B - A) + 2\varrho w' \end{aligned}$

when

$$B = b_1 + b_2 + \cdot + b_{n-1}$$
$$A = a_1 + a_2 + \cdot \cdot + a_{n-1}$$ (see p. 17.)

and η and η' are constants.

The factors μ and μ' are general and we may if we choose take them at pleasure and then seek the corresponding function.

Doing this we have μ and μ' given and also ϱ to determine $B - A$ from the relations [170].

Solving we have

$$\log \mu = 2\eta(B - A) + 2\varrho w$$
$$\log \mu' = 2\eta'(B - A) + 2\varrho w'$$

---

*) See Mittag-Leffler, Comptes rendus t. XC, 1880, p. 178.

whence

[171]
$$\eta \log \mu' - \eta' \log \mu = 2\varrho(\eta w' - \eta' w) = \varrho\pi i; \quad \left(\eta w' - \eta' w = \frac{\pi i}{2}\right)$$
$$w' \log \mu - w \log \mu = 2(B - A)(\eta w' - \eta' w) = (B - A)\pi i.$$

This solution however becomes indeterminate when $F(x)$ becomes doubly periodic, for then $\varrho = 0$ and

$$B - A = 2mw + 2m'w'.$$

This gives

$$i\pi(2mw + 2m'w') = w' \log \mu - w \log \mu'$$

whence

$$\frac{w}{w'} = \frac{\log \mu - 2i\pi m}{\log \mu' + 2i\pi m'}$$

which means that the logs of the multiplicators are proportional to their corresponding periods.

Returning to the form

$$f = \frac{\sigma(u + v)}{\sigma(u)} e^{-u\zeta v}$$

we observe that when

$$v = a_1 + a_2 + \cdots = 0$$

we have

$$v = 2mw + 2m'w'$$

and f vanishes showing that this climent can not be utilized in this case. Written as a product however and for $u = 3$ we have

[172]· · · ·
$$y = \frac{\sigma(u + a)\, \sigma(u + b)\, \sigma(u + c)}{\sigma^3 u} e^{-u(\zeta a + \zeta b + \zeta c)}$$

where

$$a + b + c = v = 0$$

and our climent may be taken as a rational function of pu and $p'u$ multiplied by a factor of the form $e^{\varrho u}$. It is moreover known that any function $f(u)$ of p and p' may be resolved in the form

$$f(u) = L + P$$

where

$$L = l_1 \zeta(u - v_1) + l_2 \zeta(u - v_2) + l_3 \zeta(u - v_3) + \cdots$$
$$P = c + \varSigma m P^{(v)}(u - v)$$

where

$$l_1 + l_2 + l_3 + \cdots = 0.$$

This property being general, we have, f being doubly periodic, but to multiply by $e^{\varrho u}$ to find a development for the climent required in [172] namely

[173]· · · · · · · · · ·
$$\Phi(u) = e^{\varrho u} \zeta(u)$$

We have then

$$\zeta(u) = \Phi(u)e^{-\varrho u}$$

$$\zeta'(u) = \Phi'(u)e^{-\varrho u} - \varrho\,\Phi(u)e^{-\varrho u}$$

$$\zeta''(u) = \Phi''(u)e^{-\varrho u} - 2\varrho\,\Phi'(u)e^{-\varrho u} + \varrho^2\,\Phi(u)e^{-\varrho u}$$

$$\zeta^{(3)}(u) = \Phi'''(u)e^{-\varrho u} - 3\varrho\,\Phi''(u)e^{-\varrho u} + 3\varrho^2\,\Phi'(u)e^{-\varrho u} - \varrho^3\,\Phi(u)e^{-\varrho u}.$$

— — — — — — — — — — — — — —

Whence

$$[174]\quad e^{\varrho u}\zeta^{(n)}(u) = \Phi^{(n)}(u) - \frac{n}{1}\varrho\,\Phi^{(n-1)}(u) + \frac{n(n-1)}{12}\varrho^2\,\Phi^{(n-2)}(u) + \cdots$$

We have then a decomposition in the form

$$[175]\quad\cdots\qquad f_1(u) = c\,e^{\varrho u} + \sum_m\sum_\nu A_{n,\nu}\,\Phi^{(\nu)}(u-\nu_n)$$

where ν_n stands for the several infinites of $f_1(u)$ and $\Phi^{(\nu)}$ for the derivatives where ν must be of an order one degree less than the multiplicity of the infinites. The coefficients A will be determined in general by developing $f_1(u)$ according to the powers of $(u-\nu_n)$ while c will be a fixed value depending upon the given conditions.

In our case then we may write

$$[176]\quad\cdots\qquad\cdots\qquad f_1(u)\,c\,e^{\varrho u} + \zeta u\cdot e^{\varrho u}.$$

This function when ν is zero, in which case $\Phi=0$ and $D=0$, takes the place of $f(u)$ and hence the general solution is

$$y_1 = f_1''(u) - 3bf_1 u$$
$$= (c\,e^{\varrho u} + \zeta u\cdot e^{\varrho u})'' - 3b(\zeta u\cdot e^{\varrho u} + c\,e^{\varrho u})$$
$$f_1'(u) = \varrho c\,e^{\varrho u} + \zeta'u\,e^{\varrho u} + \varrho\zeta(u)e^{\varrho u}$$
$$f_1''(u) = \varrho^2 c\,e^{\varrho u} + \zeta''u\,e^{\varrho u} + 2\varrho\zeta'(u)e^{\varrho u} + \varrho^2\zeta^2(u)e^{\varrho u}$$

whence

$$(\zeta u\,e^{\varrho u})'' = \zeta''u\,e^{\varrho u} + 2\varrho e^{\varrho u}\zeta u + \varrho^2 e^{\varrho u}\zeta u$$

and we have

$$[177]\quad\cdots\qquad y_1 = (\zeta u\cdot e^{\varrho u} - 3b\zeta u\,e^{\varrho u} + c'\,e^{\varrho u}$$
$$= e^{\varrho u}[\zeta''u + 2\varrho\zeta'u + (\varrho^2 - 3b)\zeta u + c].$$

But from the foregoing theory in this case we have the coefficients of $\zeta(u)$ equal to zero, i. e.

or

$$\varrho^2 - 3b = 0$$

$$[178]\quad\cdots\qquad\cdots\qquad\varrho^2 = 3b.$$

To find c we proceed as follows: --

$$\zeta u = \frac{1}{u} - \frac{g_2}{20}\frac{u^3}{3} \cdots$$

$$\zeta' u = -\frac{1}{u^2} - \frac{g_2}{20} u^2 \cdots$$

$$\zeta'' u = \frac{2}{u^3} - \frac{g_2}{10} u$$

$$e^{\varrho u} = 1 + \varrho u + \frac{\varrho^2 u^2}{2} = \frac{\varrho^3 u^3}{6} + \cdot$$

Hence

$$y = \left[1 + \varrho u + \frac{\varrho^2 u^2}{2} + \frac{\varrho^3 u^3}{6} + \cdots\right]\left\{\left[\frac{2}{u^3} - \frac{g_2}{10} u - \right]\right.$$
$$- 2\varrho\left[\frac{1}{u^2} + \frac{g^2}{20} u^2 + \cdots\right] + c + \cdots\left.\right\}$$

and taking c so that the constant term equal zero we have

[179] $\cdot$ $\cdot$ $\cdot$ $\cdot$ $\cdot$ $\cdot$ $\cdot$ $\cdot$ $c = \frac{2}{3}\varrho^3 = 2\varrho b$.

The general solution $(\nu = 0)$ is then:

$$y_1 = (\zeta u e^{\varrho u})'' - 3b(\zeta u \cdot e^{\varrho u}) + 2\varrho b e^{\varrho u}$$

where

$$\varrho = \sqrt{3b}.$$

Finis.

Table of Forms

$n = 3.$

Forms for $n = 3$.

The complete Integral is

$$y_1 = C F(u) + C' F(- u)$$

where

$$y = F(u) = f''(u) - 3 b f(u)$$

and

$$f(u) = \frac{\sigma(u + \nu)}{\sigma u \, \sigma \nu} \, e^{(x - \zeta \nu) u}$$

the ordinary form of the equation of Hermite for $n = 3$ being:

$$\frac{d^2 y}{d u^2} = [12 p(u) + B] y.$$

A second form of the integral is: —

$$y = \prod_{a = a, b, c} \frac{\sigma(u + a)}{\sigma a \, \sigma u} \, e^{- u \zeta a} = \prod_{a = a, b, c} \frac{\sigma(u - a)}{\sigma a \, \sigma \omega} \, e^{u \zeta a}$$

$$= \frac{\sigma(u - a) \, \sigma(u - b) \, \sigma(u - c)}{\sigma a \, \sigma b \, \sigma c \, (\sigma u)^3} \, e^{(\zeta a + \zeta b + \zeta c) u}$$

where

$$x = \zeta \nu - \zeta a - \zeta b - \zeta c \qquad \nu = a + b + c$$

and $B = 15 b$ which is intirely arbitrary and is originally expressed in the form

$$B = h(e_1 - e_3) - n(n + 1) e_3$$

in which case the equation of Hermite is

$$\frac{d^2 y}{d x^2} = [12 k^2 sn^2 x + h].$$

We have also the general form: —

$$y = \pm \sqrt{Y} = \sqrt{(pu - e_1)^\epsilon (pu - e_2)^{\epsilon'} (pu - e_3)^{\epsilon''}} \prod (pu - pa)$$

$$e, e', e'' = 0 \quad \text{or} \quad 1.$$

The functions developed in the general theory have values as follows:

$$\varphi = 4b^6 - bg_2 - g_3 \qquad c = \tfrac{1}{15} \qquad\qquad A_2 = \tfrac{1}{4}\,\varphi'$$

$$\varphi' = 12b^2 - - g_2 \qquad P = 15\,b \qquad\qquad A_3 = \tfrac{1}{4}\,\varphi - b\varphi'$$

$$l = 3b = \tfrac{1}{5}\,B \qquad B_0 = -\tfrac{3}{2}\,\varphi' \qquad\quad t = p\,(u)$$

$$a_1 = \tfrac{3}{4}\,g_2 \qquad\qquad B_1 = \tfrac{9}{4}\,\varphi - 6b\varphi' \quad t' = p'u = [4t^3 - tg_2 - g_3]^{1/2}$$

$$b_1 = \tfrac{27}{4}\,g_3 \qquad\qquad \gamma_1 = 0 \qquad\qquad\quad S = t - b$$

$$\varphi(t) = 4S^3 + 12\,bS^2 + (12\,b^2 - g_2)\,g + 4b^3 - bg_2 - g_3$$
$$= 4S^3 + 12\,bS^2 + \varphi'S + \varphi$$

$$S = \tfrac{1}{4}\,\varphi\,(t) - 3bS^2 - \tfrac{1}{4}\,\varphi'S - \tfrac{1}{4}\,\varphi.$$

$$Y = S^3 + A_2 S + A_3 = S^3 + \tfrac{1}{4}\,\varphi'S + \tfrac{1}{4}\,\varphi - b\varphi'$$
$$= S^3 + (3b^2 - \tfrac{1}{4}g_2)\,S - \tfrac{1}{4}(44b^3 - 3g_2 b + g_3) = \tfrac{1}{4}\,\varphi(t) - b\,(\varphi' + 3S^2)$$
$$= \tfrac{1}{4}\,\varphi\,(t) - b\,[\varphi' + 3\,(t - b)^2]$$
$$= t^3 - 3bt^2 + \left(6b^2 - \tfrac{1}{4}\,g_2\right)t - \left(15b^3 - g_2 b + \tfrac{1}{4}\,g_3\right)$$

$$Y(e_1) = -\,b\,[\varphi' + 3\,(e_1 - b)^2] = -\,b\,[15b^2 + 3e_1^2 - 6e_1 b - g_2]\;\cdot$$
$$= -\,\frac{B}{15}\left[\frac{B^2}{15} - \frac{6e_1 B}{15} + 3e_1^2 - g_2\right]$$
$$= -\,cB\,[B^2 - 6e_1 B + 45e_1^2 - 15g_2]$$
$$= -\,c^2\,Q_1 P$$

$$x^2 = \frac{125l^6 - 210a_1 l^4 - 22b_1 l^3 + 93a_1^2 l^2 + 18a_1 b_1 l + b_1^2 - 4a_1^3}{36l\,(l^2 - a_1)^2}$$
$$= \frac{4\,(l^2 - a_1)^3 + (11l^3 - 9a_1 l - b_1)^2}{36l\,(l^2 - a_1)^2}$$
$$= \frac{\Phi\,(l)}{S\,D^2}$$

where

$$\Phi\,(l) = 125l^6 - 210a_1 l^4 - 22b_1 l^3 + 93a_1^2 l^2 + 18a_1 b_1 l + b_1^3 - 4a_1^3$$

or

$$\Phi\,(\xi_1) = 125\xi^6 - 210c\xi^4 - 22\xi^3 + 93c^2\xi^2 + 18c\xi + 1 - 4c^3$$

$$S = 36l \qquad D = l^2 - a_1 \qquad \xi = b_1^{-\frac{1}{3}}\,l$$

$$c^3 = \frac{a_1^3}{b_1^2} = \frac{1}{108}\frac{g_2^3}{g_3^2} = \frac{(1 - k^2 + k^4)^3}{(1 + k^2)^2\,(2 - k^2)^2\,(1 - 2k^2)^2}$$

Also:

$$x = \frac{Q\gamma}{C B_0} = \frac{\gamma}{c^2 B_0} \sqrt{\frac{Q}{P}} = \frac{2}{3\,\varphi'} \sqrt{\frac{4\,A_2{}^3 + 2\,l\,A_3{}^2}{b}} = \frac{2}{3\,\varphi'} \left(\frac{\varDelta}{b}\right)^{\frac{1}{2}}$$

$$= \frac{1}{6\,\varphi'} \left\{ \frac{\varphi'^3 + 27\,\varphi^2 - 8\,(27)\,b\,\varphi\,\varphi' + 16\,(27)\,b^2\,\varphi'^2}{b} \right\}^{\frac{1}{2}}$$

where

$$\gamma\,Q = - (4\,A_2{}^3 = 27\,A_3{}^2) = - \varDelta \qquad C = c^4\,P\,Q_1\,Q_2\,Q_3$$

$$Q = Q_1\,Q_2\,Q_3 \qquad \varPhi\,(l) = \varPhi_1\,\varPhi_2\,\varPhi_3 \qquad y = - \frac{1}{(15)^3}$$

$$= (15)^3\,\varDelta \qquad \varDelta = \frac{1}{(15)^3}\,Q = \frac{1}{(5)^3}\,Q_1\,Q_2\,Q_3$$

$$= [4\,A_2{}^3 + 27\,A_3{}^2]$$

$$Q_\lambda = 3^2 \cdot 5\,[\varphi' + 3\,(c_\lambda - b)^2] = 5\,\varPhi_\lambda$$

$$Q_1 = B^2 - 6\,e_1\,B + 45\,e_1{}^2 - 15\,g_2 = 5\,[5\,l^2 - 2\,(k^2 - 2)\,l - 3\,k^4] = 5\,\varPhi_1$$

$$Q_2 = B^2 - 6\,e_2\,B + 45\,e_2{}^2 - 15\,g_2 = 5\,[5\,l^2 - 2\,(1 - 2\,k^2)\,l - 3] = 5\,\varPhi_2$$

$$Q_3 = B^2 - 6\,e_3\,B + 45\,e_3{}^2 - 15\,g_2 = 5\,[5\,l^2 - 2\,(1 + k^2)\,l - 3\,(1 - k^2)^2]$$

$$= 5\,\varPhi_3.$$

$$e_1 = \frac{1}{3\,\lambda}\,(2 - k^2) \qquad e_2 = \frac{1}{3\,\lambda}\,(2\,k^2 - 1) \qquad e_3 = - \frac{1}{3\,\lambda}\,(1 + k^2)$$

$$g_2 = \frac{4}{3\,\lambda}\,(1 - k^2 + k^4)$$

$$p\,(\nu) = \frac{5\,l^6 + 6\,a_1\,l - 10\,b_1\,l^3 - 3\,a_1{}^2\,l^2 + 6\,a_1\,b_1\,l + b_1{}^2 - 4\,a_1{}^3}{36\,l\,(l^2 - a_1)^2}$$

$$= k^2 sn^2 w - \frac{1 + k^2}{3} = \frac{\psi\,(l)}{36\,l\,(l^2 - a_1)^2}$$

$$= \frac{Q\gamma^2}{c^4\,P\,B_0{}^2} - \frac{2\,B_1}{B_0} - \frac{B}{2\,n - 1} = \frac{Q_\lambda\,F_\lambda{}^2}{c^2\,B^2\,P} + e_\lambda$$

$$= \frac{2160\,b^6 + 216\,b^4\,g_2 + 1080\,g_3\,b^3 - 9\,b^2\,g_2 - 54\,b\,g_2\,g_3 - g_2{}^3 + 27\,g_3{}^2}{36\,b\,(144\,b^4 - 24\,b^2\,g_2 + g_2{}^2)^2}$$

$$= \frac{\varphi'^3 + 27\,\varphi^2 - 108\,b\,\varphi\,\varphi' + 36\,b^2\,\varphi'^2}{36\,b\,\varphi'^2}$$

$$p'\,(\nu) = - k^2\,sn^2\nu\,cn^2\nu\,dn^2\nu = \frac{\chi\,(l)\,x}{18\,l\,(l^2 - a_1)}$$

$$= \frac{162\,b\,\varphi\,\varphi' - 27\,\varphi^2 - \varphi'^3}{108\,\varphi'^3\,b^{\frac{1}{2}}} \sqrt{\varphi'^3 + 27\,\varphi^2 - 216\,b\,\varphi\,\varphi' + 432\,b^2\,\varphi'^2}$$

$$= \frac{2\,F_1\,F_2\,F_3}{c^3\,P\,B_0{}^3} \sqrt{\frac{Q}{P}}$$

$$pv - b = \frac{\varphi'^3 + 27\,\varphi^2 - 108\,\varphi\,\varphi'}{36\,\varphi'^2 b}$$

$$pv - e_\lambda = \frac{Q_\lambda F_\lambda^2}{c^2 B_0^2 P}$$

$$= \frac{[\varphi' + 3\,(e_\lambda - b)^2]\,[12\,(b - e_\lambda)\,(2b - e_\lambda) - \varphi'^2]^2}{36\,\varphi'^2 b}$$

$$\frac{p'v}{2x} - pv = b - \frac{3}{2}\frac{\varphi}{\varphi'} \cdot$$

where

$$\psi(l) = \Phi(l) - 12\,l\,(l^2 - a_1)\,(10\,l^3 - 8a_1 l - b_1)$$
$$= 5l^6 + 6a_1 l - 10b_1 l^3 - 3a_1^2 l^2 + 6a_1 b_1 l + b_1^2 - 4a_1^3$$
$$\chi(l) = \tfrac{1}{2}[\Phi(l) - 3\psi(l) - 108\,l^2(l^2 - a_1)^2$$
$$= l^6 - 6a_1 l^4 + 4b_1 l^3 - 3a_1^2 l - b_1^2 + 4a_1^3 = A \cdot B \cdot C$$
$$A = l^2 - (1 + k^2)\,l - 3k^2 = \tfrac{45}{2}F_1$$
$$B = l^2 - (1 - 2k^2)\,l + 3\,(k^2 - k^4) = \tfrac{45}{2}F_2$$
$$C = l^2 - (k^2 - 2)\,l - 3\,(1 - k^2) = \tfrac{45}{2}F_3$$
$$F = F_1 F_2 F_3 = \frac{8}{3^6 5^3}\chi = \frac{8}{3^6 5^3}A \cdot B \cdot C.$$

Case 1. $P = 0$.

Integral a special function of Lamé of the first sort.
$$y = p'. \qquad B = 0.$$

Case 2. $Q = 0$; $\Phi(l) = 0$; $Q_1 = 0$; $Q_2 = 0$; $Q_3 = 0$; $x = 0$;
$p'v = 0$; $v = w_\lambda$.

Integrals, six in number, of the second sort.

$$f = \pm\frac{\sigma_\lambda u}{\sigma u} = \frac{\sigma(u + w_\lambda)}{\sigma u\,\sigma(w_\lambda)}\,e^{-u\,\zeta(w_\lambda)} \qquad \begin{array}{l} a = 1,\,2,\,3 \\ \zeta(w_\lambda) = \eta_\lambda. \end{array}$$

$$= z\,\sqrt{pu - e_\alpha}$$

where

$$z = pu - \tfrac{1}{2}e_\alpha - \tfrac{1}{10}B$$

(a) $Q_1 = 0$

$$B = 3e_1 + \sqrt{3\,(-12\,e_1^2 + 5g_2)} = k^2 - 2 \pm \sqrt{(k^2 - 2)^2 + 15k^4}$$
$$y = \left\{p + \tfrac{1}{2}e_1 - \tfrac{1}{10}\left(3e_1 + \sqrt{3\,(5g_2 - 12e_1^2)}\right)\right\}\sqrt{p - e_1}$$
$$= \left\{p + \tfrac{1}{15}\,(k^2 - 2) \pm \tfrac{1}{10}\sqrt{(k^2 - 2)^2 + 15k^4}\right\}\sqrt{p - \tfrac{1}{3}(k^2 - 2)}.$$

(b) $Q_2 = 0$

$$B = 3e_2 \pm \sqrt{3(5g_2 - 12e_2)} = 1 - 2k^2 \pm \sqrt{(1 - 2k^2) + 15}$$

$$y = \{p + \tfrac{1}{2}e_2 - \tfrac{1}{10}(3e_2 \pm \sqrt{3(5g_2 - 12e_2^2)})\}\sqrt{p - e_2}$$

$$= \{p + \tfrac{1}{15}(1 - 2k^2) \pm \tfrac{1}{10}\sqrt{(1 - 2k^2)^2 + 15}\}\sqrt{p - \tfrac{1}{3}(1 - 2k^2)}$$

(c) $Q_3 = 0$

$$B = 3e_3 \pm \sqrt{3(5g_2 - 12e_3^2)} = 1 + k^2 \pm 2\sqrt{(2 - k^2)^2 - 3k}$$

$$y = \{p + \tfrac{1}{2}e_3 - \tfrac{1}{10}(3e_3 \pm \sqrt{3(5g_2 - 12e_3^2)})\}\sqrt{p - e_3}$$

$$= \{p + \tfrac{1}{15}(1 + 2k^2) \pm \tfrac{1}{5}\sqrt{(2 - k^2)^2 - 3k}\}\sqrt{p - \tfrac{1}{3}(1 + k^2)}.$$

Case 3.

$$F_\lambda = 0; \quad \chi = 0; \quad A = 0; \quad B = 0; \quad C = 0; \quad \nu = \omega_\lambda;$$

$$x \neq 0$$

$$f = -\frac{\sigma_\lambda u}{\sigma u} e^{x u} = \frac{\sigma(u - \omega_\lambda)}{\sigma u} e^{(x - \zeta(\omega_\lambda)u)}.$$

Six values of this form corresponding to the roots of $A = 0$; $B = 0$; $C = 0$, namely

$$B = \tfrac{5}{2}(1 + k^2) \pm \tfrac{5}{2}\sqrt{(1 + k^2)^2 + 6k^2}$$

or

$$B = \tfrac{5}{2}(1 - 2k^2) \pm \tfrac{5}{2}\sqrt{(1 - 2k^2) - 6(k^2 - k^4)}$$

or

$$B = \tfrac{5}{2}(k^2 - 2) \pm \tfrac{5}{2}\sqrt{(k^2 - 2) + 6(1 - k^2)}$$

which determine corresponding values for x.

Case 4. Conditions as in case (3) with the additional condition of the functions of M. Mittag-Leffler.

The integral is:

$$y_1 = (\zeta u e^{\varrho u})'' - 3b(\zeta u e^{\varrho u}) + 2\varrho b e^{\varrho u}$$

where

$$\varrho = \sqrt{3b}$$

$$\Phi = -6A_2 S^2 + 9A_3 S - A_2^2$$

$$E = -9[2A_2 S - 3A_3]$$

$$\Psi = \gamma Q.$$

CPSIA information can be obtained
at www.ICGtesting.com
Printed in the USA
BVHW04*1046170918
527708BV00015B/1895/P